Exemplary College Science Teaching

Exemplary College Science Teaching

Edited by Robert E. Yager

Arlington, Virginia

Claire Reinburg, Director
Jennifer Horak, Managing Editor
J. Andrew Cooke, Senior Editor
Amanda O'Brien, Associate Editor
Wendy Rubin, Associate Editor
Amy America, Book Acquisitions Coordinator

Art and Design
Will Thomas Jr., Director

Printing and Production
Catherine Lorrain, Director
Nguyet Tran, Assistant Production Manager

National Science Teachers Association
David L. Evans, Executive Director
David Beacom, Publisher

1840 Wilson Blvd., Arlington, VA 22201
www.nsta.org/store
For customer service inquiries, please call 800-277-5300.

16 15 14 13 4 3 2 1

NSTA is committed to publishing quality materials that promote the best in inquiry-based science education. However, conditions of actual use may vary and the safety procedures and practices described in this book are intended to serve only as a guide. Additional precautionary measures may be required. NSTA and the author(s) do not warrant or represent that the procedure and practices in this book meet any safety code or standard or federal, state, or local regulations. NSTA and the author(s) disclaim any liability for personal injury or damage to property arising out of or relating to the use of this book including any recommendations, instructions, or materials contained therein.

Library of Congress Cataloging-in-Publication Data
Exemplary college science teaching / edited by Robert E. Yager.
pages cm
Includes bibliographical references and index.
ISBN 978-1-938946-09-7
1. Science—Study and teaching (Higher) I. Yager, Robert Eugene, 1930- editor of compilation.
Q181.E938 2013
507.1'1—dc23
2013016540

Cataloging-in-Publication Data are also available from the Library of Congress for the e-book.
e-ISBN: 978-1-938946-80-6

Contents

Preface

Dr. Brian R. Shmaefsky
President, Society for College Science Teachers

Professor of Biology, Lone Star College, Kingwood, Texas

Recognizing exemplary practices in the college science classroom was the primary motivation for the establishment of the Society for College Science Teachers (SCST) on March 24, 1979, in Atlanta, Georgia, during an NSTA national conference. The founders of SCST wanted an organization dedicated to the improvement in the teaching of college science courses through interdisciplinary collaboration between teachers of college science. In April, 1981, SCST became an official affiliate of NSTA and started providing college-level NSTA members increased services to help them to reach their personal objectives as well as those of the profession.

Monographs such as "Exemplary College Science Teaching" represent one way NSTA and SCST collaborate to serve college instructors. It truly reflects SCST's interdisciplinary approach to studying and promoting the advancement of college science teaching. This monograph is dedicated to the community of college and university teaching scholars who are working to enhance science education through the development and testing of best classroom teaching practices.

Recently, I came across an excerpt from an unlikely reference while surfing the internet for information about the history of college science teaching. It was in a book called *The Scottish Connection: The Rise of English Literary Study in Early America*, written by Franklin E. Court in 2001. Court described the criteria for best practices in a college education for 1803 as the ability to recite hundreds of pages of facts and rules.

My further investigations into past science education literature showed this to be the norm from 1803 to the 1920s. Critical analyses of college education by educational researchers steadily grew from the 1920s to the 1950s. By the 1940s, educational researchers were downplaying the value of rote memorization in science teaching. The ideas of John Dewey fueled a new way of instruction called progressive education (Westbrook 1991).

It appears that Dewey, and other education reformers, were more effective at communicating their ideology to the K–12 educators. College teachers were encouraged to teach content in an environment that fostered exploratory learning and relevance. Plus, they applied rational assessment models as formulated in Benjamin Bloom's classification of educational objectives (1956). The infrastructure of the reform movement had the following elements:

- Lead the student to become aware of a problem (or gap in information to resolve an issue).
- Direct the student to define the problem.
- Have the student formulate rational hypotheses to resolve the problem.
- Lead the student to evaluate the consequences of the different hypotheses based on supported evidence.
- Encourage the student to test the likeliest solution typically as a group discussion.

This "innovative" instructional methodology showed measurable successes at improving the quality of education and led to the funding of many progressive science teaching strategies by the National Science Foundation. It also gained support by most state education departments and was becoming the prevailing educational model of the United States Department of Education.

Unfortunately, higher education did not adopt progressive educational strategies until the new educational paradigm was established in K–12 schools. The model of college rote content area "expertise" was considered the prevailing formula for success. Student success is currently the "buzzword" in college science teaching; accreditation and funding entities now look for systematic pedagogical strategies that foster student success in college. Each manuscript in this monograph is derived from the contemporary need to improve college science education. Plus, many of the authors are SCST members who long ago were promoting best practices in college science teaching.

As you read this monograph, think about the following questions about your teaching. What type of instructional model are you using in your classroom? Is it contributing to student success in discipline area and does it inculcate lifelong skills? How do you learn about new trends in science education and do they really improve college science teaching? What is your role to society as a science educator? Is your instructional strategy promoting workforce skills in the sciences? Is your department and is your administration supportive of excellence in teaching? Does your college or university encourage the explorations of novel teaching strategies?

References

Bloom, B. S. 1956. *Taxonomy of educational objectives.* Boston: Allyn and Bacon.

Westbrook, R. B. 1991. *John Dewey and American democracy*. Ithaca, NY: Cornell University Press.

Foreword

Exemplary College Science Teaching: Helping or Hindering STEM Reforms?

Robert E. Yager
University of Iowa

Problems With Most College Science Teaching

Finally we come to grips with the cause for needed reforms of science education that include the college level! For some reason few have questioned the teaching of college science courses and how it models teaching for those who aspire to teach in K–12 schools. Elementary education students frequently do not feel interested, successful, or positive with their performances in science classes they have experienced in high school or college. Yet the attitude and creativity found in students in elementary schools result in most positive views regarding science learning! Too many future teachers in middle and high schools often merely mimic what they experienced in college. Ironically, the lack of experience with college science results in more positive attitudes about "doing" science at the elementary level, especially when teachers admit to not knowing!

The worst problem in terms of school science is the efforts at the upper high school levels, where only half of the students complete any science courses, often in chemistry and physics. These two offerings at 11th and 12th grades are too often labeled as college preparation courses, where the assumption is that the teaching should model what characterizes these disciplines in colleges.

College science teaching usually does not focus on how the content would be used for various courses in secondary schools. There is rarely any attempt to focus on student learning and how teaching and the curriculum might be affected by college instructors. In fact, most were concerned that most grade 5–12 science teachers had not completed enough college coursework to qualify them to teach science. Major NSF funding for science teachers often focused on completing more coursework in science as a way to improve. There was little or no focus on *how* science was done or for understanding how "doing" real science could help in resolving personal and societal problems. "Doing" science goes beyond what is done or what most researchers do in their own laboratories. Paul DeHart Hurd noted in 1980 that science had become primarily a collaborative undertaking concerning how the natural world could be studied. Hurd reported that one effort published in France included more pages of the names of scientists who contributed than pages actually devoted to report the research findings.

College teaching is too often offered as only something done by speaking (lecturing) to students about the ideas and theories describing the universe in discipline formats, as is commonly found in courses in physics, chemistry, and biology. In addition to lecturing, most college scientists include some descriptions of their experiences in the laboratories. These experiences are meant to serve as ways of portraying the science done by scientists, and offer other views concerning relationships of "learning" what was presented in lectures. There is often little or no "hands-on" learning science itself or any use of the materials presented. In many ways college science is offered as what current scientists accept as accurate interpretations of the features of nature as described by "experts," sometimes with their supporting evidence (proof). There is often no action to encourage students to think and do other than remember what they were told (like following a recipe in cooking)!

It is only recently that college teaching has been analyzed and alternative suggestions tried. This is what the Exemplary Science Program (ESP) encourages all to consider. The results concerning such thoughts and ideas that have been tried are reported in the 16 chapters of this monograph at a time when new reforms are being considered (STEM).

The first chapter is offered as the experience and pathway of a science teacher questioning the typical repeating of what was experienced in college as a teaching model. It sets the stage for the many changes needed and undertaken—probably by many of the authors of the following chapters. For reforms to succeed, many more trials, features, and questions must be a focus. The hope is that this monograph and the 16 exemplars will enlarge, encourage, and support changes in college science teaching!

Engaging Students in Actually "Doing" Science

Research about student learning highlights the fact that real learning is focused on thinking, analyzing, and personally constructing meaning. Gee (2011) identified several researchers in the late 1950s and early 1960s that helped develop the constructivist theory of learning. These researchers include Bruner, von Glasersfeld, Gergen, Piaget, and Vygotsky. The aspects of constructivist theory have been offered as variations, including radical constructivism, social constructivism, and social cultural constructivism. In general, constructivists believe that learners construct knowledge rather than merely receiving it; and this act of construction is greatly dependent on the prior knowledge and experience that the learner brings to the task. Von Glasersfeld (1995) explains that from the constructivist perspective, learning is not a stimulus response phenomenon. It requires self-regulation and the building of conceptual structures through reflection and abstraction. Generally constructivist practices consist of

1. posing problems of emerging relevance to learners or expect students to question themselves;
2. structuring learning around "big ideas" or primary concepts;
3. seeking and valuing students' points of view;
4. adapting curriculum to address student suppositions; and
5. assessing student learning in the context of teaching.

Students should be encouraged to start with problems and questions. Science itself begins with questions, followed by encouraging and expecting others to identify multiple forms of evidence that might offer answers. Students are expected to consider as many answers as possible. These can be compared among students in a given classroom. Next is a discussion of ways of testing the accuracy of the ideas generally. Again, unique results provide settings for more student examples. Perhaps many of them are never mentioned previously or in all class sections.

It is important for teachers to reject the idea that science is something that can be transmitted directly to students. Too often teachers only expect students to repeat what textbooks or teachers say. This can be done in ways like they are the parts from a school class play. The more students are involved, the more they are like the actors mirroring "doing" science better. For learning to succeed, students must work to understand and use their efforts in order for them to indicate success with learning experiences. Most teachers, however, are willing to accept verbatim definitions students can remember from teacher statements or books as verifications. What happens when results of lab activities are given? They are too often simply explanations from a lab manual. Many of the chapters included in this monograph report on the inclusion of community members who can help with actually engaging students in the processes of science itself.

Local Issues and Happenings

Too often the materials included in a lecture, in a directive lab, or as ways of assessing science learning are not related to any real-life use. Community involvement is rarely related to anything beyond what comprises a lecture, outside assignment, or a lab. Science teachers should be instrumental in noting actions and happenings that affect the whole community, that respond to specific problems, or that illustrate ties to a given issue. Too often the curriculum, content covered, and laboratories are unrelated to what students can use—especially in new situations. Effective college teaching particularly must relate to the content and procedures found in communities, where both teachers and students work. Application of lecture explanations and lab activities should provide new contexts that illustrate real learning.

Some teachers like to introduce a problem with immediate ties to a planned laboratory, relating such labs to local businesses, environmental issues, and/or recent local concerns (new reports). Too often the stated problem is a statement given by teachers that offers little impact or interest for all students. Again, students can help! There are differences among students in science classes. Teachers should take advantage of such differences among students by sharing excitement, ideas, problems, probable conclusions, and actions as ways of promoting student thinking and learning!

Teachers use plans, teaching topics, textbooks, and laboratories that too often do not provide and enhance planning, interest, action, or use in the specific communities. Encouraging students to compare activities and actions should occur frequently. Colleges have an advantage of drawing more students into classes than what can be tried in K–12 communities. Teachers need to be aware of the diversity represented, all of which can inspire better learning. It sometimes adds uniqueness for a course when the teacher asks students to consider why specific problems and issues are included. Who thinks that the topics comprising labs and lectures are perceived as important? Too often students are not part of the leadership; they often do not see themselves as more than recipients of what teachers expect them to do themselves!

No Lectures

A successful college teacher needs to be willing to appreciate that lectures represent teacher actions from which little real learning results. This monograph includes many ideas and suggestions of why and how faculty procedures are sure to be unsuccessful when lectures are expected and offered as the main focus for teaching.

It is important for teachers to ask what the goal is for a given course. Too often this is not even considered. In fact in many high schools, courses are structured around specific concepts just because it is assumed that it is just what they are expected to do! Too often college teaching does not include student participation, interactions, and experiences. Again, the major focus should be on student learning. Too often the lecture method is what is expected by the department head or by the university leaders as a whole. Too many faculty members from research universities are quick to indicate that they are primarily interested in their own research, publications, and success as being more important than teaching a course. Teaching is something they are expected to do—to share their understandings with students. The quick and easy way is to prepare a lecture, probably like the ones they experienced as students themselves. Most are not willing to admit that their own learning did not come from a lecture. Many never give it a further thought. They rarely see teaching as a major effort related to their role as a professor. It is rare for the most successful students to be able to do more than to repeat what they "learned" with the same explanations and visions that were used by the professor. Again, such teaching is an act of transmission! Few science professors have ever had a pedagogy course or even heard of constructivist learning theory. A reform of college teaching is needed to attract more students to careers, to develop better citizens, and to find desired employment.

Technological Aids

Current society sees advantages of technological advances for education. One can use technology to enhance typical teaching (i.e., continuing to encourage students to accept the explanations that accumulate from the work of scientists). Technology is often not understood nor in any way controlled. But it often is used as related to lectures (and associated labs). It often characterizes and relates to structured curricula. Sometimes college teachers are forced to use lecturing to handle larger numbers of students or to get specific information to illustrate industries associated and found at other locations and environments.

Some of the uses of technology are tied to educational technology. Some of our reviewers were critical and questioning about technology as examples of what is appropriate for exemplary teaching. The nice part of technology is the engagement of students in planning and using it as a tool for learning. Most of the examples and improvements have only loose ties to a given community. It is also of little use to merely use technology to supplement the science lectures and cookbook labs.

The current reforms often include emphasizing technology (the T of STEM), which is critical as we redefine content and teaching goals for all K–16 grades. It would be interesting to see where STEM efforts in all states produce changes that relate to improving the teaching outcomes that most reforms seek to produce.

Science Begins With Problems: Whose Problems? Society, Scientists, or Students?

Few college science classrooms start with student questions or issues. How does this affect college teaching, not focusing on students, their experiences, ideas, and needs? Too few see ways to change teachers who teach major parts of their own major discipline area. Too seldom do college teachers start as scientists, separating "doing" science from merely choosing facets of "agreed upon" content included in textbooks.

Most standardized test providers (who are central to college science teaching) do not claim to know the meaning of science, or how it should be used in teaching. Testing experts prefer to perceive what teachers teach as enough and all they need to do to prepare test material to match what occurs in *their* classrooms and laboratories. Tests should reflect real learning by students, not merely what they remember.

Reforms are difficult! But, it is important to realize that learning is an individual experience. Sometimes the best learning occurs from errors. Scientists need to analyze their science experiences and what research unveils about learning. Everyone (including technology experts) need to then prepare exams that focus on improved learning for students, for colleges, and for themselves. Sometimes this will involve actual use of the ideas and skills in a new context. Some call for more teacher education for college science faculty, saying that college science teachers must know science for their science careers (including a special line of research). This too often ignores the needed role for exemplary teaching!

Service Learning

One of the chapters (chapter 14) deals with service learning. While not something that all should do, service learning offers important possibilities. Questions can be resolved, community members can become involved, and changes can be made. Of importance is how improved teaching is related to all levels of high school and college teaching of science.

The important aspect is whether all students are involved and if all the needed "services" can revise teaching that places the teacher as part of societal improvement. The advantages include a way of meeting some of the other aspects of current reform efforts. These include defining a problem, community involvement, including all the STEM areas, with actual results in teaching STEM and the needed associated research. Such efforts illustrate the power and effort gained by collaboration among teachers, students, and others in a given community.

One could argue, however, that service learning would be a way to illustrate real teaching improvement as a definition while also helping decide what information should be available for student work. It is also a way to assess teacher experiences while also ridding science of merely recasting what fairly typical science has unfortunately accomplished.

College science instructors often find it hard to relate to what the National Science Education Standards for K–12 teaching advocates. Too often students never know how science is practiced. Service learning provides new parameters for changing thinking and the ways students are introduced to it. It promotes a vehicle and replacement of the course lectures and associated labs. Again, these teachers choose what are often not found as meeting student needs and interests. It is hard to use constructivist theory when a college teacher has never experienced it him- or herself.

Changing Teaching Versus More of the Same

Change is difficult! But, change is always necessary if real reforms are to occur. These chapters provide important ideas and results to promote science and specific changes for many who see the light; but, how to encourage more change? This is the primary purpose of the NSTA ESP effort.

It was Paul Brandwein who often spoke of how to get more involved with the reforms and how to get their experiences shared and accomplished to advance the rest of society. We hope that all the authors involved with the college teaching monograph will continue to use, change, and encourage others to share ideas. It is like becoming a scientist in an education situation. Again, this means starting with a problem, preparing possible solutions, gathering evidence that it works, considering what others have done to keep learning continuous. We look forward to publicizing the successes and "spreading the word." This is considered the major focus for this monograph.

We hope to remain in touch as more changes are tried to offer new successes that are experienced. We continue to analyze the debates. It is hard to imagine what will happen as new students are taught using the ways envisioned. Maybe new features of constructivist theory will be more central in changing teaching in colleges. We hope the new features of teaching are more successful with greater learning results and speed up the process.

Science Is Basic for Learning in General

With successful science teaching, it is easy to see ties to journalism, all the successful STEM features (especially with respect to mathematics), art, and the humanities. All associated issues for improving education are enhanced by science; activities and real science experiences in colleges should be available (Saul et al. 2012). The current reform movement centers on major changes as set forth with the science as contained in textbooks and curricular frameworks. It is instead something where the successful educators want changed perspectives and use of constructive theory, just as important theories influence science.

There is evidence that every student can be involved using his or her own experiences with science. This does not mean all need to do the same things—learning occurs ideally when experienced in diverse ways. There is no quick fix when teachers change their teaching strategies. The change should focus on each one involved with experiments in teaching (i.e., learning in constructivist ways, which means that teachers become more engaged). Learning is then seen as more successful, and science itself is seen as more interesting, meaningful, and useful. Many students never accept science as more than encouragement to take the next course for a college major. Why is the encouragement needed to inspire and encourage more for college science teaching? Change in college teaching is what is needed to improve life as we know it. The reforms described in this monograph illustrate what can be done.

The Power of Differences Among Students

Many teachers set up barriers for learning when starting with their basic plans, textbooks, and their own personal interests. Too often students are caught up in doing what their teachers expect of them. Student creativity is something students experience; it is natural—but too often it is not valued or encouraged in educational settings. Student-centeredness is something most will subscribe to, but not something nurtured or used to interest students. Too many ignore students, especially in classes enrolling 50–500 students. Students are all different, but too often their differences are not used to promote learning. Nor is student-centeredness practiced as a way of inviting teams of students to work on individual or group projects. If reforms are desired, they must be original and conceived by students—not merely something students volunteer to do and to conceive as their own. The focus on proposed projects should come from groups of students, not teacher lesson plans. Students can be more successful later in encouraging other students to be involved if they experience such teaching themselves. Changes should mirror work in communities among various groups comprising a society.

Focus on Learning Versus Teaching

College teaching should be concerned first of all with student learning and not what college faculty members find interesting or what they are particularly knowledgeable about when they work with assigned learners. But, typical science teaching in college environments too often ignores student interests and past experiences. Students are supposed to absorb what teachers present to others to earn grades, indicating success in most college courses. Typical science teaching should be sharing a common "story." It should be a reflection of what each student becomes. It should be related to their work as students and include their own research. Effective teaching means being personal and enthusiastic!

Teaching rarely veers from what has been "common," with the generally accepted view that it is the teacher's course to plan and carry out. Students too often are but "recipients." If science teachers can allow students to "do" science, there will be more learning, more positive student attitudes, and more connections to a community of learners and citizens in general.

Unfortunately teaching is not considered as something to test. Instead, most college courses tend to be structured to provide specific information and skills for students to "learn," with the end result that they merely memorize concepts. If education is an activity that can be studied and improved, we will have a revolution concerning real learning—especially where one is caused by college science faculty.

The 16 stories in this monograph are all fine examples of how college teaching must change if stated objectives and experiences are to result in successful learning. Readers are encouraged to read and react to each concerning their own teaching efforts. There are likely great differences and some similarities. This set of examples and the ESP monograph are offered to encourage more analyses and current experimental efforts concerning exemplary college teaching. They represent well what STEM efforts are designed to promote!

References

Cullen, R., M. Harris, and R. R. Hill. 2012. *The learner-centered curriculum: Design and implementation.* San Francisco, CA: Jossey-Bass.

Gardner, H. 2008. *Five minds for the future.* Boston: Harvard Business Press.

Gee, J. 2011. *Beyond mindless progressivism.* Available online at *www.jamespaulgee.com/node/51*

Saul, W., A. Kohnen, A. Newman, and L. Pearce. 2012. *Front-page science: Engaging teens in science literacy.* Arlington, VA: NSTA Press.

Von Glasersfeld, E. 1995. Constructivist approaches to science teaching. In *Constructivism in education*, ed. L. P. Steffe and J. Gale, 3–15. Hillsdale, NJ: Erlbaum.

Acknowledgments

Members of the National Advisory Board for the Exemplary Science Series

Shelley A. Lee
Science Education Consultant
WI Dept. of Public Instruction
P.O. Box 7842
Madison, WI 53707-7841

Lisa Martin-Hansen
Associate Professor of
Science Education
College of Education
P.O. Box 3978
Georgia State University
Atlanta, GA 30302-3978

LaMoine L. Motz
Science Facilities and Curriculum
Consultant
8805 El Dorado Drive
Waterford, MI 48386-3409

Edward P. Ortleb
Science Consultant/Author
5663 Pernod Avenue
St. Louis, MO 63139

Carolyn F. Randolph
Science Education Consultant
14 Crescent Lake Court
Blythewood, South Carolina 29016

Barbara Woodworth Saigo
Science Education and Writing
Consultant
President, Saiwood Publications
23051 County Road 75
St. Cloud, MN 56301

Pat M. Shane
NSTA President 2009–2010
Clinical Professor and
Associate Director (Retired)
University of North Carolina at
Chapel Hill
Center for Mathematics and
Science Education
Chapel Hill, NC 27514-6110

Patricia Simmons
Professor and Department Head
Math Science & Technology
Education
North Carolina State University
Raleigh, NC 27695

Gerald Skoog
Texas Tech University
College of Education
15th and Boston
Lubbock, TX 79409-1071

Mary Ann Mullinnix
Assistant Editor
University of Iowa
Iowa City, Iowa 52242

About the Editor

Robert E. Yager

Robert E. Yager—an active contributor to the development of the National Science Education Standards—has devoted his life to teaching, writing, and advocating on behalf of science education worldwide. Having started his career as a high school science teacher, he has been a professor of science education at the University of Iowa since 1956. He has also served as president of seven national organizations, including NSTA, and has been involved in teacher education in Japan, Korea, Taiwan, Indonesia, Turkey, Egypt, and several European countries. Among his many publications are several NSTA books, including *Focus on Excellence* and two issues of *What Research Says to the Science Teacher.* He has written more than 700 research and policy publications as well as having served as editor for eight volumes of NSTA's Exemplary Science Programs (ESP). Yager earned a bachelor's degree in biology from the University of Northern Iowa and master's and doctoral degrees in plant physiology from the University of Iowa.

The Road to Becoming an Exemplary College Science Teacher

Katherine B. Follette
University of Arizona

Setting

I aspire to be an exemplary college science teacher. I entered graduate school to study astronomy with this as a primary personal goal for my education, equal to, if not even surpassing the more common goal of establishing myself as a research scientist. Having just come from a brief but rewarding career as a high school and middle school math and science instructor, I was prepared to put in my time and spend six (or more) years improving my research and teaching credentials such that I could get a job at an institution of higher education that emphasizes both.

I was not well prepared for how miserable it would make me. I had two clear goals for graduate school, but after two semesters, I didn't feel that I had made progress toward either. Upon returning from a spring break trip to India, I burst into tears to find myself back in my cubicle facing another round of problem sets that I couldn't do and scientific papers that I couldn't decipher.

If I am the hero of my own story, then this point is what Joseph Campbell called "the belly of the whale" in my journey. This was the moment when I almost gave up on what I had set out to do upon embarking on my graduate school quest. I either needed to change course and make my way out of the darkness, or risk being broken down by the gastric juices of graduate school.

With the benefit of hindsight, I know that what I was experiencing is called "imposter syndrome" and that it is very common, particularly among women in the sciences. I even have some thoughts about what we as a community should do about it. However, this is not a book about graduate education reform; it is a book about college teaching.

It was a focus on that first goal of becoming an exemplary college teacher that saved me from dropping out of graduate school, pulled me out of my deep dark cave of self-pity, put my situation and my science back into perspective, and reminded me of why I wanted to be an astronomer in the first place. My salvation came in the form of an e-mail that underwent multiple forwards before alighting in my inbox: a call for applicants to teach an introductory astronomy class at the local community college.

It is as a community college instructor and a (still enrolled) graduate student in the sciences that I write this, reflecting on my professional development experiences and what I learned from each. I emphasize that all of these were "extracurricular" in the sense that I did them in my "free time," which often meant in lieu of sleeping or attempting to achieve work-life balance. I hope that it will serve as a justification of sorts for why all graduate students should be encouraged to pursue such opportunities as an essential part of their academic career.

Motivation

Whether or not they ever find themselves at the front of a college classroom, I find it hard to imagine that any scientists can make it through graduate school and into stable careers without at some point being expected or required to communicate what they do and why it is important to superiors, colleagues, politicians, governing bodies, or the public. I have had the privilege of meeting many successful scientists thus far in my career, and one thing that I have noticed about all of them is that, beyond their frequently lauded scholarly abilities, they tend to be remarkable communicators. In fact, I would go so far as to say that I have never encountered a truly spectacularly successful scientist who was not a good communicator as well. This is why the development of science communication skills should be an essential step in the making and minting of any modern scientist.

The first time I was encouraged to consider teaching as a career was in college, but it didn't come from a science professor. It came from a Japanese language instructor, and I remember being surprised and even a little insulted. We rarely hear scientists speak of teaching as a legitimate career option. Teaching was for "burnouts" and people who "just didn't cut it" doing legitimate science. Even writing these words, I can feel my hackles rising, but it is important to keep this attitude in mind when discussing how to become an exemplary college science teacher, as it is something that we need to overcome, individually and as a community.

Of all of the students I have taught at the secondary and postsecondary level, I remember best those who told me that they were inspired to go into teaching or astronomy after taking my class. We delight in making clones of ourselves, and if I am being completely self-aware, I should recognize this as both the reason I remember these students and also the fundamental reason I was never directly encouraged to go into teaching by a science instructor.

Even though many of the instructors from whom I draw my greatest inspiration are college science teachers, I believe that they would identify themselves first and foremost as "physicists," "astronomers," and "planetary scientists." To say that they are "college professors" perhaps doesn't capture the fact that they are also in many cases practicing research scientists—at least not to the general public—but to say that they are "scientists" alone *also* doesn't capture the other aspects of their jobs, at least not under the current connotation of the word.

This speaks to the purpose of this chapter in two senses. First, we could do better as a community at acknowledging that very few of us do "just science" or "just teaching." The majority of academics in the sciences regularly do both. Secondly, the lack of encouragement that I received and the length of time it took me to stumble upon teaching as something that suited my disposition and my talents demonstrates that professional development is just as important for college science teachers as it is for secondary school educators.

If my science instructors had recognized their teaching as an important and valued part of their professional identity (and indeed many of them were quite good at it), perhaps I would have received such encouragement from them rather than from instructors in other disciplines. If I had realized earlier that my desire to be a scientist and my desire to communicate science were compatible, perhaps I would have seen these instructors as role models for how to do science *and* how to communicate it.

Skills such as recognizing our own biases and learning to overcome them are rarely innate and are in some cases contrary to basic human nature, but they are important to develop in order to become an exemplary college science teacher. We are unlikely to encounter such information organically, so emphasis on professional development by both academic departments and individual instructors is essential. If we are going to acknowledge the dual roles of educator and researcher espoused in the phrase *college science professor*, then we should also acknowledge that both roles need to be fostered, encouraged, and developed. The purpose of professional development is to inform us of issues in modern education, results of the vast body of educational research that few of us can hope to monitor in addition to our other responsibilities, and to encourage awareness and skill development.

This chapter is not meant to be a how-to; I neither expect nor recommend that anyone take the same convoluted path toward becoming a college science instructor that I have. Rather, this chronicle is meant to be an encouragement to think about the role that educational professional development experiences can (and should) play in the making of a young scientist, whether he or she intends to go into "pure" research, "pure" teaching, or some impure combination of both.

Classes in Pedagogy

Why aren't pedagogical classes a required component of every graduate education? Most people will say something along the lines of "because not all scientists need to teach." This is baloney. All scientists have to communicate, very few of us do it well, and the claim that scientists do not need teaching skills completely ignores the fact that graduate departments already emphasize broad education, at least as it applies to their discipline. For example, although I study the formation of planets in our own galaxy, three of my eight required graduate courses focused on extragalactic phenomena.

I will probably never apply for time on a telescope to study anything outside of our galaxy, but I would never argue that these classes were worthless. First of all, very few of us have the luxury of teaching only within the narrow subfield of our discipline that is the focus of our research. Many of us will be asked to teach survey courses or skills courses, and the smaller the institution we end up at, the more likely we are to have to teach on a topic that is entirely unrelated to our particular subfield.

Secondly, science today is tremendously interdisciplinary and thinking about approaches and techniques used in other disciplines or subdisciplines can spark ideas for new approaches and techniques in our own. I am always telling my students to look at the "big picture" and to see the connections between things. Not just between the variety of topics covered in the curriculum, but between the science they are learning in my classroom, the material they've learned in their other courses and the information that they encounter in everyday life.

Even scientists who never teach formally in a classroom must write grants that justify the "broader impacts" of their work and have to interact with friends, family, and the people sitting next to them on airplanes who ask: "And what do you do?" Even those who leave academia to pursue careers in industry could benefit from pedagogical training in the sense that they will need to communicate their ideas, conclusions, and expertise to colleagues with very different backgrounds. We should all be able to explain what we do and why it is interesting and important, and recognize and confront misconceptions as they arise.

For example, when I tell people that I study the formation of planets around other stars, it usually sparks a conversation about aliens. They often ask whether I believe they exist, and I say "yes, absolutely I believe that in all of the planets around all of the stars in all of the galaxies of the universe there is probably even another astronomer out there answering this very question right now, but..."

That "but" is very important. It is only through my pedagogical training and classroom experience that I know that the first part of my statement will lead to misconceptions if I do not explain it further. Because I regularly think about the distances between stars, I know that a galaxy "teeming" with life would mean that the nearest intelligent, communicating civilization would be a journey of at least several years away traveling at the speed of light, which is not (yet) possible. However, for the average person, discussion of the probability of alien life, which most astrobiologists will agree is very high, summons images of all of their favorite science fiction characters traversing the galaxy at "warp speed" in exploration, trade, or conquest. They picture a galaxy teeming with life in the same way that a coral reef or a rain forest or even a drop of pond water teems with life.

In terms of communicating to them the tremendous scales between objects in our universe (and how truly remarkable it is that we know anything about them at all), it is probably just as important that I have had the experience of interacting with students and therefore know the common misconceptions about the probability of intergalactic travel and meaningful alien communication as that I have a basic understanding of the fundamental laws that make such rapid travel physically impossible.

Similarly, a thorough understanding of the geometry of spacetime surrounding a spinning black hole is not sufficient to guarantee success in explaining to a student, a layperson, or even a scientist in a different discipline, what a black hole is and why it is important that we know about them. If a scientist is not aware of jargon and is unable to read and engage non-scientists in such conversations, then he or she will probably end up doing more harm than good in shaping the opinion of science (and scientists) held by students and people encountered by chance in the course of everyday life.

We scientists do a terrible job of communicating to the public what we do and why it plays an important role in society. We are no longer living in the Apollo heyday, when John F. Kennedy was able to come right out and say how much the space program was costing each citizen of the United States per week with confidence that public opinion would still be behind him. As modern society becomes increasingly complex and modern science becomes increasingly esoteric—as we are no longer engaged in an outright race with another country to achieve a scientific goal—the public will find it increasingly difficult to get behind science funding and our ability to communicate will become more and more important to the survival of our profession.

In my class, one of the first exercises that I have my students do is to define the terms *science, math, astronomy,* and *astrology*. The word "difficult" comes up frequently, as do phrases like "a lot of facts" and "not my thing." Astronomy and astrology are frequently seen as synonyms. Active verbs are almost never used.

What this tells me is that my community college students, who come from a wide range of backgrounds and represent a reasonable cross section of the population, do not know what science is or value it as important in their lives. I just recently decided to test this directly by asking them on their first homework assignment whether and how they think science is "important" in their lives. Most answers affirmed that science was important, I am sure at least partially because they think that is what I want to hear, but when challenged to come up with an example of how science affects their everyday lives, the responses were generally vague.

If my students mentioned any specific science, it was usually medicine. They never mentioned how medicine is related to science, or that it is fundamentally biology and chemistry research that we mean when we say "medical science." They also sometimes reference technology (particularly cell phones), but again in a vague sense and without tying it to any specific scientific discipline or advancement.

Not a single person mentioned any phenomenon or concept that they would have learned about in an introductory science course. None mentioned the NSF, NASA, NIH, or any scientific organization or group. Not only does the general public not know *who* does science, they can't articulate *why* it is important in any more than a vague way. Yet we expect that when the next round of budget cuts come around, they will stomp their feet and declare "No! Don't cut science. Science is *important*!"

If the average citizen does not know what science is or why it is important, that is our fault. It is the fault of every scientist who has ever spoken to a school group, relative, politician, or layperson and not noticed the glazed eyes, stunned expressions, or outright boredom of their audience. Effective communication of science concepts, whether it be in the classroom or on the airplane, is an important skill to foster in every scientist, regardless of their intended career trajectory. I have a hard time seeing how a couple of required classes on pedagogy could hurt even the most dedicated lab scientist in their career.

Pedagogical training has helped me to give better talks to my colleagues, not only to my students, and to be better able to read whether audiences are following me so that I can adapt, slow down where appropriate, and clarify where needed. How many of us have sat through a talk by a brilliant scientist on a topic that is interesting and relevant to our own work but found ourselves watching the clock, falling asleep, playing with our phones, or counting the nose hairs of the presenter? How frequently have we become lost by jargon or been in need of clarification during scientific presentations? One benefit of pedagogical training is that it should help you identify the weaknesses in your own presentation skills and teach you how to read your audience, whether that audience is your students or your peers. Understanding how people learn, and particularly how *other* people learn, is a revolutionary experience that can carry over to how you communicate with everyone.

Before I took a class on pedagogy, for example, I assumed that most learners were fundamentally like myself. I am a compulsive overachiever who enjoys traditional lecture classes. I have always hated group work. I taught myself biology by reading a textbook because when I took my Biology 101 course I was also working full-time and rarely made it to class. I had heard of "visual learners" and "auditory learners" and "learning styles," and I even understood in principle that people learn a little differently from one another.

However, I also assumed that everyone could learn as much as they needed to from a proper lecture or from reading a good textbook if they really applied themselves. It was not until I was confronted with the tremendous body of research saying otherwise that I believed this to be wrong, and not until I reflected on my incorrect belief that my own education had been guided primarily by lecture and textbook-reading that I understood how wrong my assumption had been.

What did I truly learn from my undergraduate education in physics, for example? I remember some great classes and great lectures about some very interesting things, but what actually stuck with me long term was a way of approaching a problem based on prior knowledge, mathematical reasoning and an understanding of how the world works. I did not learn that in class by watching my professors work through problems on the board, no matter how well I followed their reasoning at each step. I learned it later, when I had to go back through my notes and use them as a tool to answer new questions that were not in the textbook and were unlike those that were covered in class. I learned most of what I still know today by debating the answers and methodology to new problems with my classmates.

I did not learn to think scientifically by watching demonstrations or reading textbooks or even listening to skilled lecturers. I learned it by conducting experiments, considering whether my answers made sense, and reexamining when they did not. I did it by debating methodology and correctness with someone else, and that someone was more frequently a peer than a professor.

Even coming to this realization did not sell me on the philosophy of "interactive learning" that so much of modern pedagogy focuses on. If I did most of my learning outside of the actual classroom, isn't that where everyone else should do it too? Isn't it a "waste" of class time to engage in interactive learning when I should be covering more content? If I have to devote two classes to a topic that would otherwise take only half a class in order to engage my students in interactive learning, then aren't I doing them a disservice by covering less content than I would otherwise?

If you have made it to the point of teaching or pursuing a degree in a particular discipline, then you should realize that you are *not* an exemplar of the typical learner in your field. It may seem to you that to choose between teaching your students about the nature of light or the laws of gravity, but not both, is like having to choose between teaching your five-year-old to tie her shoes or blow her nose into a tissue instead of her sleeve. They need both to survive, don't they? In the case of science content knowledge, the answer is a resounding no.

Let's assume, for example, that I was not able to get to gravity in my introductory astronomy class because we spent our time talking about motions in the solar system in a general sense and how they can help to explain phenomena like the yearly motion of the sun or the cause of a lunar eclipse. If I have done my job, and my students are really comfortable with the spatial reasoning skills needed to explain these motions, then if someone tells them that there is a 9th(or 10th) planet known to ancient civilizations "hiding" behind the Sun that will reappear in 2012, impact the Earth and cause the apocalypse, my students should immediately recognize this as false. They do not need to be able to refute that claim by applying the laws of gravity. In fact, if they failed to remember much about rotation and revolution and the predictable motions of objects in the solar system, which I presumably covered in lieu of gravity, I would not be so terribly disappointed either. They could take any number of reasonable scientific approaches to refuting that claim and that is what I *really* want them to leave my course with—the ability to *reason* scientifically.

We forget because we find our own subjects (and often the sound of our own voices) so fascinating that the purpose of education is to develop a set of skills that can be applied in broad contexts. When we spend all of our own time studying the same subject, we lose sight of this, and lose sight of our purpose as instructors. We fail in our duty to teach all of our students by ignoring the needs of those whose interests and inclinations do not align with our own. These students may never become scientists, but they can still benefit from what we have to teach them, and they can still incorporate scientific reasoning into their worldview.

We have a great deal of power to affect public perception of science and the level of scientific literacy in our country by reaching the nonmajors in our classrooms. Even in classes for majors and graduate students, we could take much more time to focus on the broad applicability of the skills we are emphasizing in all disciplines and subdisciplines rather than focusing on content knowledge alone. We need pedagogical training to be able to do this well.

Without question, incorporating interactive learning into your classroom practice is difficult. Not only does it require you to cover a smaller number of topics in your course, it does not always go smoothly the first time you try it.

The first semester that I tried using lecture tutorials in my classroom, I had a student complain to the dean of the college that I was having them "do worksheets" and was refusing to lecture for the entire class period like a proper instructor. It was my first semester teaching a college class, just as I was making my way out of the deep dark cave of misery that had been my first year of graduate school, and I was devastated.

Luckily, I had just completed a pedagogical class focused primarily on the usefulness of lecture tutorials in the classroom. It was only because I had so recently seen the data suggesting that this technique could work that I didn't give up on it right there and then. I had to assume that just as I had not believed my own education was driven by interactive learning, this student probably held the mistaken belief that he would learn a lot more if I lectured from the beginning to the end of class (2.5 hours) like I was "supposed to." I was able to articulate this to the dean (using my communication skills) and to back up my claims with data, like any good scientist should. The data also gave me the confidence I needed to keep at it. At the end of the semester, that very student validated my claim by begrudgingly acknowledging that some of the lecture tutorials were "all right."

I was lucky. I had a mentor close at hand who was able to point me to the appropriate data in the educational literature, translate the jargon, and explain the conventions of educational statistics so that I could formulate a sound argument in response to this student complaint. Most of my colleagues, most senior professors even, would not have been so lucky. Just as the jargon, conventions and unspoken rules unique to our scientific disciplines would be daunting for a novice trying to familiarize herself with it by perusing the literature, one can't expect to get a sense for the huge body of educational research simply by reading a couple of articles in the *Chronicle of Higher Education*, which is as far as pedagogical training goes for many faculty. This is why pedagogical classes, taught by discipline and broader education specialists, can be so useful. They can boil down the huge body of educational research to a handful of points that are particularly relevant to our discipline and give us the tools we need to keep up on some of the relevant literature in the future.

Since that first semester of teaching with lecture tutorials, I have become better at "selling" them to my students, and I haven't had any further complaints. However, since I teach in my "free time" and my efforts are not considered part of my degree program, there would be very few consequences if I were to get a few. If I were freshly on the tenure track and felt that every move I made was being scrutinized, on the other hand, that one complaint to the dean would have been enough to turn me away from interactive learning forever.

This is why pedagogical classes alone cannot make good college science teachers. The ideas learned in the pedagogy classroom have to be put into practice in your own classroom, and often have to be attempted several times, before they sink in and become a part of your practice. Thus, actual teaching also needs to be a component in the development of college science teachers, and if we want the college science classroom to reflect modern pedagogical techniques, this training needs to come into play *before* the junior professorship.

The Teaching Assistantship

The one piece of a typical graduate curriculum that involves teaching is the Teaching Assistantship. Of course, the amount of teaching that one does in a Teaching Assistantship varies widely from school to school and professor to professor. Some TAs teach large portions of a class, some teach only discussion or problem sessions, some simply hold office hours and some do not interact with students at all, but serve as a grading lackeys. In my experience, the majority fall into the latter categories rather than the former, particularly in the sciences.

In fact, it seems that a large number of new faculty step in front of a classroom for the first time having received very little practical training from their two or three graduate Teaching Assistantships. For the instructor and his or her students both, this is something like a children's swim class in which the instructor was hired simply because they themselves know how to swim. Even an Olympic swimmer would have trouble with this task without the proper preliminary training. If we simply throw them in the pool with the students and see how they do, it is not that hard to imagine having several students who are so thoroughly traumatized by the experience that they never set foot in a pool again. Nor is it hard to envision that the instructor might have no desire to repeat the experience, though they may have to grin and bear future requests to do so.

Not only do we scar teachers and students alike under this model, but it is completely contrary to the principles of good teaching. On the first day of class, would you hand your students the final exam and wish them luck? It is little wonder that so few college science faculty consider their teaching to be part of their identity. They are set up to fail, and nobody wants to claim as a part of their job description something they are terrible at, which is precisely what we should expect most of them to be in this "sink or swim" model of professional development.

Aside from exceptionally gifted natural communicators, none of us are exemplary teachers our first time around. The TAship should be a perfect opportunity to ease people into the experience and to offer mentorship and constructive feedback. Every TA should be required to teach at least one full class during their semester-long TAship. Every instructor who wishes to have a TA should agree to this on day one. Rather than having a TA "cover" for them when they are out of town or very busy, the faculty member should be present for these classes in order to provide feedback and encouragement. Grading papers is not going to teach you how to be a good instructor, and indeed we call these positions *teaching* assistantships because they are supposed to involve *teaching*.

The Education community expends a great deal of effort advocating for a move away from the traditional "learning by osmosis" paradigm, but then we turn around and train our graduate students to teach by precisely that methodology. Sitting in the back of a classroom and policing cell phone usage, setting up demos, proctoring exams and grading papers are not what a graduate *teaching* assistantship should be about.

Among the most beneficial professional development experiences I have had as a college science instructor were peer and self observations. These experiences were not part of a teaching assistantship, but a Certificate in College Teaching program that I pursued in my "free time." Because I was teaching my own community college class, I was able to have a number of people come and observe me and provide feedback and constructive criticism. I also had several classes videotaped and was asked to provide my own feedback. Watching myself teach was painful, but also revolutionary.

My teaching today is much better for having had these experiences. As evidence that improvement happens organically given practice and feedback, consider Figure 1 (p. 10), which shows the improvement in my own teaching evaluations over the course of three semesters teaching the same class.

We need a paradigm shift in the field of teaching assistantships. Both students and departments should take them more seriously, and they should be seen as an integral part of a graduate curriculum rather than a required departmental service. They should have a curriculum of their own that involves feedback from peers and professors, and they should be accompanied or preceded by pedagogical training. This is the only way that we can expect to develop a large body of exemplary college science instructors—by training them to teach and then giving them an opportunity to practice it in a friendly and encouraging pre–tenuretrack environment.

I lay out a sample curriculum for a pedagogical course for graduate students at the end of this chapter (see Appendix, p. 13). As a pure thought experiment, this curriculum, like all teaching products, could certainly be improved through implementation and feedback, and I hope that I get the opportunity to do so in the future.

Continued Professional Development

Teaching people how to communicate should not end with graduate school. When I was a high school teacher, I was required to do a certain number of professional development hours per year. Why should this be any different for college faculty? There are opportunities in abundance, but not much incentive for faculty to participate.

Figure 1.1. Results of Student Course Evaluations From My First Three Semesters of Teaching at the Community College Level.

All three were the same course—"Stars, Galaxies, Universe." Response options were a Likert scale from "Strongly Agree" to "Strongly Disagree." "Not Applicable" was also an option, but has only been included in the charts in instances when at least one student selected it. It is worthwhile to note that while these were my first three semesters of teaching at the college level, I already had two years of experience teaching at the secondary level when I began. Had this truly been my first time in front of a classroom, the improvement may have been even more dramatic and the starting level would likely have been lower.

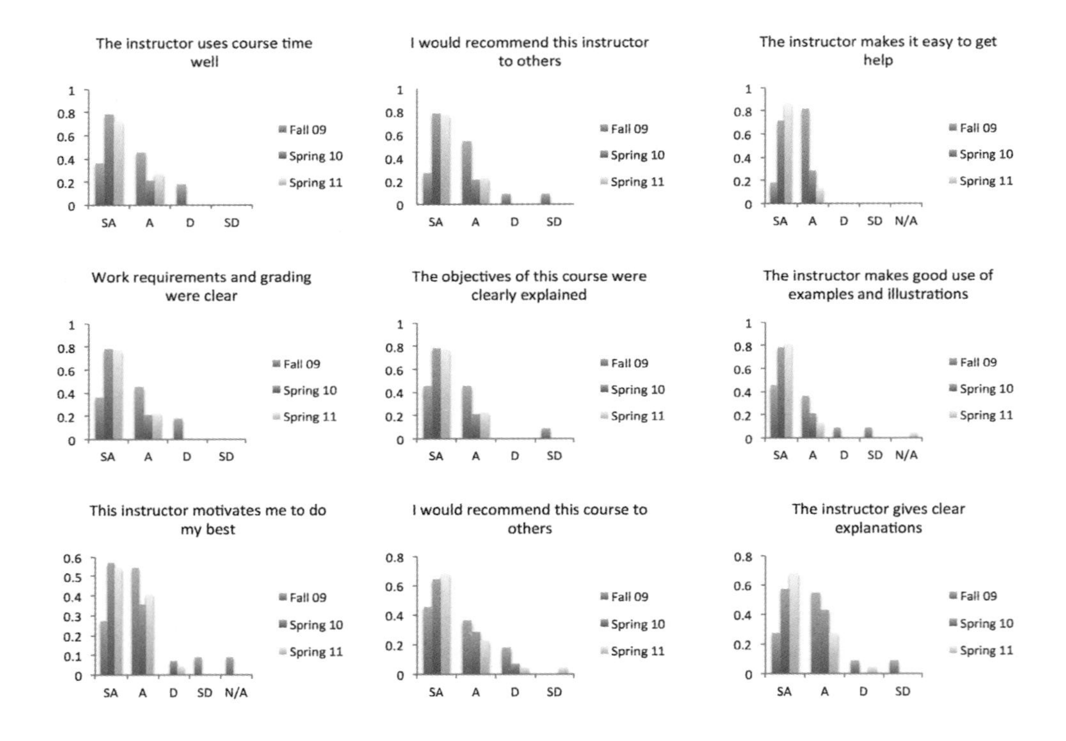

As a community college instructor, I have attended a number of workshops that focused on a certain pedagogical tool or technique, ranging in length from one hour to two days. Like me, the other faculty attending these workshops were generally there on their own time and without the direct encouragement or financial support of their departments. I have begun to recognize their faces (and make a few friends) because it is the same small group at these workshops every time. Most took place before or during scientific conferences, where presumably more than half of the conference attendees have some teaching responsibilities, so where are the rest of the faculty?

If I were one of these other faculty, I am not sure that I would be interested. If there was no incentive or requirement for me to participate and I already had the overfull schedule of the typical college science faculty, I probably wouldn't make the time to attend. If my teaching evaluations had little to no bearing on my tenure application, then I would most likely choose to make a new connection that might blossom into a joint science paper, which would help my case, rather than electing to learn a new pedagogical technique, which would not.

I believe that this is the biggest barrier to the proliferation of exemplary teaching faculty—we do not value it enough as a community to make it a priority all the way from graduate education to tenure and beyond. We have few standards and virtually no requirements. We consider teaching ability to be a "bonus" skill, something that some of our colleagues have that we admire them for, but is not essential for success. This is a delusion. The most successful scientists are almost always good communicators. As it stands today, I can only assume that they achieve these skills through luck, giftedness or perseverance. If we want them to be more widespread, we must make communication and pedagogical training part of the professional development of every scientist.

Many of my suggestions along those lines, I realize, involve additional requirements and responsibilities, and the schedule of the typical college science faculty, postdoc, or graduate student is already overfull, so where is the time supposed to come from? We can't simply tack on additional requirements and expect faculty to further compromise work-life balance. I know that I, for one, in trying to fulfill both the research requirements of my department and my own personal goal of improving my teaching abilities, have often felt that my own work-life balance and personal relationships were compromised. This should not be the only way to become good at both. If my husband and I had children now, it wouldn't even be a possibility for me to continue like this.

Much like incorporating learner-centered teaching techniques into your own classroom, professional development is not free and requires a critical look at what is truly important, a cutting of the fat at every level.

Maybe the conscious incorporation of pedagogical training into graduate education means that students are required to take one fewer science course or write one fewer science paper in order to graduate.

Maybe for junior faculty it means that they are expected to do one fewer departmental service or teach one fewer class in order to focus on teaching *well*.

Maybe for senior faculty this means taking on some of these departmental services and finding the time to mentor graduate students and junior colleagues in teaching skills, not just research techniques.

What we need most urgently is simply a change in attitude. If the community could be made to value communication skills as much as more traditional scientific skills, then I believe that our classrooms, our professional lives, and the opinion of the general public toward science would be much improved.

It is hard to quantify benefits like these, therefore programs that aim to improve science communication seem to be the first on the chopping block in our current economy, as evidenced by the recent cancellation of the National Science Foundation's GK12 Program and the 25% cut to NASA's education budget, to name just two instances that I am aware of.

If you are reading this book, then you are probably already one of the converted, so I will leave you with the idea that we can *all* do more, even those of us who are already thinking about these problems. If you are a junior scientist or above, make time to improve your communication skills and to engage other scientists in dialogue about its importance. Give good lectures and science talks, interact with your education colleagues, encourage feedback from students, peers and superiors, do outreach when you can, and take the time to think about the clarity of your scientific and educational messages when you speak to administrators and members of the public. If you are a midcareer scientist or above, demonstrate to those you mentor that you value communication skills, and encourage them to pursue their development. Allow them to invest a portion of their time in the development of these skills and give them constructive feedback. Do not treat these activities as extracurricular for you or for them. If you are a senior scientist, support your colleagues in these endeavors and also keep in mind that you have the power to change department culture and policy at a high level.

A grassroots effort to change the culture of science could begin with all of us making efforts to improve the dialogue in our own classrooms and departments. If the effort were to spread from there, it could ultimately lead to a marked improvement in the understanding and perception of science in the world at large. That may very well be the most valuable contribution that we as scientists can make to society, and it is exemplified in the making of exemplary college science teachers.

Appendix 1.1

Outline for a One-Semester Class on Pedagogy for Graduate Students

Suggested Format

1, 2, and 3 credit options
2.5 hours
One meeting per week
12 weeks
Class size: 10–20

Basic Daily Schedule

45 min–1 hr 15 min: Lecture/Presentation (by instructor and/or expert in topic)
Break
1 hr–1.5 hr: Activity/Practicum

Course Description

This course will provide a broad overview of 10 major topics in teaching and learning, pedagogy, student engagement, classroom management, and presentation skills. Students will put the skills they develop into practice through class activities.

Course Objective

Students become familiar with several major concepts in teaching and learning and develop their understanding such that they know how to seek out appropriate resources for more information and continue to improve their own teaching and communication skills in the future. Students establish a cohort within their department with whom they can consult in the future.

Homework

All enrollees are expected to prepare for the in-class activities. Those taking the course for 2 or 3 credits will be responsible for two to four readings on pedagogy per week. Students will write one-page responses to the readings before class in an online discussion forum, and will comment on the responses of two of their classmates. Those taking the course for 3 credits will also be expected to complete three teaching observations and reflections over the course of the semester, and will compose a teaching philosophy.

Week	Topic	Lecture/Discussion Topics	Activities/Practicum	Objectives
1	Introduction	What qualifies as teaching? What qualifies as "good" teaching? Teaching styles Introduction to the body of literature on teaching and learning	Each student is asked to reflect on the qualities/teaching styles/techniques of: The best class and/or instructor they ever had in their own discipline The worst class and/or instructor they ever had in their own discipline The best class and/or instructor that they ever had in another discipline The worst class and/or instructor they ever had in another discipline Students form small groups, share their lists of qualities with one another and discuss/compile. Entire class debriefs and makes a master list of qualities of good/bad instructors/courses.	Students understand that there are a variety of teaching styles, but also qualities that good teachers share. Students begin to reflect on their own teaching styles.
2	Learning	Learning styles Inductive and deductive reasoning Interactive learning Learning objectives	Students are asked to consider the same four classes they discussed last class—How much content do they recall? What else did they learn? How do they use knowledge gained in that course in their work? How do they use it in their daily life? This reflection is used for an entire class discussion of the principles of learning.	Students understand the basic research on learning styles.
3	Effective Communication	Jargon and the Retroencabulator (YouTube) Tips for using PowerPoint effectively Top tips from top communicators	Each student comes prepared with a "Fame Lab" style presentation: 3 minutes to communicate a concept of their choosing, no PowerPoint or printed materials. Guest judges and classmates provide feedback.	Students reflect on qualities of good communicators and make plans to improve their own communication styles. Students communicate a science concept without slides or excessive props.

Week	Topic	Lecture/Discussion Topics	Activities/Practicum	Objectives
4	Tutoring/One-on-one Interaction	Recognizing, probing and correcting misconceptions *Private Universe* Leading questions How to help without "giving away" the answer	Divide class into "students" and "instructors" and give each group the same lecture tutorial or homework set. "Students" decide what misconception(s) they are likely to have and "do" the homework according to these misconceptions. "Instructors" discuss solutions, alternative answers and likely misconceptions and brainstorm leading questions. Pair one "instructor" with one "student" and do a mock office-hour session. Full class debrief: Were "instructors" able to spot the misconceptions? Were "students" able to behave like real students?	Student build their ability to recognize misconceptions and develop strategies to mitigate them Students practice spotting and correcting misconceptions Students engage one another in mock tutoring
5	Lecturing	Storytelling Student Engagement/ interactive learning Techniques TPS/Clicker Questions Lecture Tutorials What to do when you "run out of time"	Students break into groups of 2–3. Each group chooses a lecture topic and designs a 10–15minute lecture (with engagement) on the topic, including slides. Students present their lecture strategies to the class.	Students are introduced to and practice several specific interactive learning techniques Students begin to contemplate and practice lecture design
6	Classroom Management	Challenging classroom situations (plagiarism, lack of motivation) Laying out classroom policies Consequences and discipline	Class brainstorms types of challenging classroom situations, how to spot them, and how to manage them effectively. Each class member is assigned one of these problems to act out. Each group from last class "gives" their prepared lecture (alternating who plays the role of instructor) and practices. management techniques as students act out their assigned roles Class debriefs/compiles a list of tips and tricks for classroom management.	Students develop skills in spotting challenging classroom situations and devise solutions for classroom management

Week	Topic	Lecture/Discussion Topics	Activities/Practicum	Objectives
7	Effective Labs and Demonstrations	Designing demos and budgeting time Integrating lecture and lab What to do with malfunctions and unexpected results	Students come prepared with a demonstration of their favorite concept (either one of their own design or one that they've seen before) and lead the class through it. After each demonstration, class provides feedback: Might it elicit any misconceptions? Might anything harm your ability to do this demo (classroom size or setup, class size, weather)?	Students practice giving classroom demonstrations Students discuss and consider how to integrate demonstrations into other content
8	Accessibility	Learning Disabilities Physical Handicaps Technological tools	As students come into class, assign each of them a "disability." Have them act this out or consider it during the lecture/reading discussion portion of the class, which is purposely designed to be inaccessible. Students with physical disabilities and those with learning disabilities form groups and redesign the class, then present their solutions.	Students reflect on common varieties of learning difficulties and develop tools to recognize them Students develop skills to make their classroom more accessible
9	Assessment	Writing effective questions Using and designing rubrics	Students come prepared with an example of a "good" and a "bad" test or homework question and class discusses what makes the good ones good and bad ones bad. Groups of 2–3 students choose a topic and design a mini lecture tutorial to test student understanding, which they present to the class during the debrief. Class provides feedback on the lecture tutorials developed by their peers. Students are assigned the task of developing a rubric for Presentation Day for next class.	Students diagnose test questions according to the principles of good assessment Students reflect on how to measure/gauge the effectiveness of assessments and design one of their own

Week	Topic	Lecture/Discussion Topics	Activities/Practicum	Objectives
10	Designing a Course	Defining Scope and Sequence Tying to course objectives	Students come prepared with an outline for their "dream" course (the course that they most want to teach). In pairs, students review and revise their course outline according to topics discussed during first portion of class. Class debriefs the activity and discusses difficulties and tips for success. Students share the rubrics they prepared for the class presentations (next class) with a group of 2-3 classmates. Each group prepares two criteria for the official rubric to be used next class. Class discusses the criteria prepared by each group, combines/modifies as appropriate, and arrives at a final rubric.	Students practice designing a course Students discuss the difficulties in determining the scope of a course and how to tie content to course objectives Students design their own rubric for class presentations next class
11	Presentation Day		Students come prepared with a 10–15 minute mock class on a topic of their choice and present it to their classmates. Each classmate and instructor(s) fill out the class-designed rubric and provide comments for each presenter.	Students practice teaching!
12	Wrap-up/Self-Assessment	Learning from student evaluations The teaching journal How to talk about teaching with your peers Positive reinforcement and constructive criticism	Students pair up and watch the video of themselves and their partner presenting last class. After each video, both students take about 5 minutes to reflect on how they or their partner could improve and debrief with one another. Class as a whole debriefs: Was it hard to watch yourself Was it hard to provide good feedback to your partner?	Students reflect on resources for continued improvement in their teaching/communication skills

Lecture-Free College Science Teaching: A Learning Partnership*

Bonnie S. Wood
University of Maine at Presque Isle

Setting

Lecturing is the predominant pedagogy in postsecondary education, and the University of Maine at Presque Isle (UMPI) is no exception. One of seven autonomous campuses within the University of Maine System, UMPI offers undergraduate liberal arts and selected professional programs to both traditionally and nontraditionally aged students. The 150-acre campus is surrounded by the rolling hills and potato fields of rural Northern Maine, but is within the city limits of Presque Isle with its 9,500 residents.

Seventy-nine percent of UMPI's 1,500 students are from Maine and the majority of those attended high school in Aroostook County, the northernmost county in Maine, bordered by the Canadian provinces of New Brunswick and Quebec. Although the county (6,453 square miles) is larger than Connecticut and Rhode Island combined, the population is only 76,000 people residing in two cities, 54 towns, 11 plantations, and 108 unorganized townships. Presque Isle is the primary commercial center of the region and is 150 miles north of Bangor (population about 35,000) and almost 300 miles north of Maine's largest city, Portland (population about 66,000).

What UMPI students lack in racial and ethnic diversity they compensate for in range of academic and social preparation for college. A typical science class roster includes a mixture of traditionally aged and much older students. Science majors and nonmajors are not separated into separate sections. Professors face the challenge of providing a rigorous undergraduate education while giving every student in their classrooms the opportunity to succeed. Most science educators learned successfully via the lecture method and we tend to model our own teaching on the way we were taught; but the lecture method is ineffective for many of today's students.

---

* Adapted from Wood, B. S. 2009. *Lecture-free teaching: A learning partnership between science educators and their students.* Arlington, VA: NSTA Press.

Overview of the Program

The authors of the *National Science Education Standards* (NSES) recognized the hurdles encountered by today's science educators by suggesting the following changes in emphases for teaching standards (NRC 1996, p. 52). My design for "Lecture-Free Teaching" supports each of these changes.

Less Emphasis on

Treating all students alike and responding to the group as a whole
Rigidly following curriculum
Focusing on student acquisition of information
Presenting scientific knowledge through lecture, text, and demonstration
Asking for recitation of acquired knowledge
Testing students for factual information at the end of the unit or chapter
Maintaining responsibility and authority
Supporting competition
Working alone

More Emphasis on

Understanding and responding to individual student's interests, strengths, experiences, and needs
Selecting and adapting curriculum
Focusing on student understanding and use of scientific knowledge, ideas, and inquiry processes
Guiding students in active and extended scientific inquiry
Providing opportunities for scientific discussion and debate among students
Continuously assessing student understanding
Sharing responsibility for learning with students
Supporting a classroom community with cooperation, shared responsibility, and respect
Working with other teachers to enhance the science program

Evidence for Success

Over the past two decades, evidence against the success of the lecture method has mounted and many consider it ineffective for teaching students of any age (Mazur 2009; Handelsman et al. 2004). Disadvantages of lecturing can be summarized in five broad categories:

- Most lectures are too long for effective learning. Attention increases from the beginning of the lecture to ten minutes into it and then progressively decreases.
- Lectures do not promote long-term retention of information and understanding of concepts. Students retain minimal understanding of content after completing a lecture-based course and lectures do not effectively rid students of previous misconceptions.
- Lectures cannot teach learning, thinking, and other behavioral skills. Lectures are not well suited to developing higher-order cognitive skills such as analysis, synthesis, and evaluation, all of which are more important than content for success after college.
- The lecture gives the teacher too much control, with little flexibility to respond to student feedback. A lecture presumes that all students enter the classroom with the same level of understanding and are learning at the same pace.

- The lecture came into being to impart knowledge. Technology, having placed knowledge at one's fingertips, has rendered the lecture obsolete.

A common thread running through all research comparing lecturing to other pedagogies is that when instructors lecture, students are passive. The consensus is that the lecture has a place, but used alone is rarely adequate to achieve learning goals. Lectures should be used only if blended with other instructional methods. I contend that if the instructor so desires, the formal lecture can be abandoned altogether and replaced with thoughtful use of active-learning strategies in a collaborative, cooperative, and supportive classroom environment in which the students and the instructor construct a "partnership of learning" to explore ways to apply knowledge.

For more than two decades I did what most postsecondary science educators do: I taught as I had been taught. Using a style modeled after my own college and graduate school experiences, I gave three 50-minute lectures each week along with a three-hour laboratory for each course. Sometimes different professors taught the lecture and laboratory portions of a course. I accumulated a large collection of transparencies and annotated my lecture notes to display transparencies at appropriate moments. I punctuated my lectures with humorous statements I had conceived over the years and updated my notes to reflect current research. At each class meeting I moved smoothly and energetically through the scheduled topics while my students furiously scribbled notes. I knew they were attentive because they asked probing questions, such as "Could you repeat that please?" or "Will this be on the test?" My goal was to perfect my lectures.

Although my students wrote flattering evaluations at the end of each semester, I began to realize that neither the weaker nor the stronger students could reliably apply what they had learned. Over the next decade I gradually reformed all of my science courses until I had abandoned lectures and replaced them with what I termed "Lecture-Free Teaching." Improving introductory science courses is my focus, but my new methods have naturally spilled over into all my upper level courses.

I began restructuring my teaching by engaging in some kind of active learning exercise during every class meeting. I initially adapted ideas gleaned from books and journals, but as I became more comfortable with my new approach, I designed my own activities. Next I increased the time allotted to each class meeting. After incorporating activities into my classes, students began to complain that 50-minute class periods were too short. Both my students and I felt rushed and the topics lacked coherence when activities ended abruptly or had to be completed at home or during the next class. I changed class meetings from 50 minutes three times a week to 75 minutes twice a week.

Ultimately, I erased the arbitrary boundary between "lecture" and laboratory. In contrast to the traditional weekly schedule of three 50-minute lectures plus a 3-hour laboratory at a different time of the week, my classes now meet in the laboratory twice a week for two approximately 3-hour sessions. I gradually replaced instructor demonstrations and cookbook-style laboratory exercises with inquiry-based activities. In cooperative learning teams, students apply the scientific process to develop hypotheses and then design and perform experiments. To complete the circle of reform, I investigated methods of formative and summative assessment and created classroom assessments to complement and reinforce my active-learning and inquiry-based approaches. On end-of semester course evaluations students now comment how much they like the two longer class meetings each week compared to the usual traditional format they experience in science courses taught by others.

Many books and articles guided my journey to lecture-free teaching. Other publications, discovered late in my reform process, demonstrate how those educators and I uncovered similar flaws in the traditional pedagogy, traveled different routes to transform our teaching, but ended up with solutions that share many characteristics (Fink 2003; McManus 2005; Wiggins and McTighe 2006). My steps for course design, as well as theirs, are adaptable to teaching in a wide range of disciplines and course levels and also embody the NSES.

Next Steps

My book, entitled *Lecture-Free Teaching: A Learning Partnership Between Science Educators and Their Students* (Wood 2009b) is a detailed description of my strategies for lecture-free teaching as well as a discussion of research supporting the need for science education reform. My 13 steps to lecture-free teaching embody the entire process of course design and delineate a never-ending feedback loop providing information to both students and instructor. For one semester my students and I are members of a "learning partnership," which contributes to our mutual goal of significant learning that remains meaningful beyond the final exam. The steps, described briefly below, are explained with examples in my book in Chapter 3, The Chronology of Course Design.

Step 1: Consider the Unique Situation of the Course You Are Preparing to Teach

Whether you are planning changes in a class you have previously taught, or designing new curriculum, there are always conditions you cannot alter. In *Creating Significant Learning Experiences: An Integrated Approach to Designing College Courses*, Fink (2003) lists six situational factors to consider when planning a course.

The first situational factor is "specific context of the teaching/learning situation." This category includes the number of students enrolled in the class; whether the course is for high school students in a particular grade or whether it is an introductory, upper level, or graduate course at a college or university; the length and frequency of the class meetings; the time of day and days of the week for class meetings; whether the classroom is a large lecture hall, a smaller classroom, a laboratory, or a seminar room; and the type and arrangement of seats and tables within that room.

A hallmark of lecture-free teaching is flexibility, and I try to be adaptable when responding to the factors above. However, as I gain experience and confidence teaching outside the lecture mode, I become more adept at persuading colleagues and administrators that changes in the length and frequency of my class meetings, as well as the arrangement of the classroom seats, enhance student learning. If we ever build a new classroom building, I will lobby for flexible seating to facilitate communication among students working in cooperative learning groups. But factors that cannot be changed will always exist and must be considered when designing a course.

Fink next describes "expectations of external groups." At the college level, an institution may dictate course or curriculum learning expectations, as might a professional accreditation organization. The catalog description of the course, often written by several educators who teach the same course, must be taken into account. When designing my courses, I consider how to meet these expectations in creative ways with my lecture-free pedagogy, remembering that science content can be delivered by a variety of methods.

Third is the "nature of the subject." Science courses are typically categorized as a combination of cognitive and acquisition of physical skills—primarily working toward a single right answer—and nearly always dynamic, with multiple opportunities to learn about changes and controversies. These characteristics can be used to a science educator's advantage in a lecture-free format.

Fourth are "characteristics of the learners" that include the current life situations of the students; the personal or professional goals they have for the course and their reasons for enrolling; their prior experiences, knowledge, skills and attitudes about the subject; and, importantly, their learning styles. This situational factor is challenging for me because of the heterogeneity of learners at my institution. I strive to present a course that is rigorous but at the same time accessible to students with poor preparation and study skills. Accommodating dissimilar students can be challenging, but the variety inherent in lecture-free teaching improves my chances of creating an appropriate learning environment for all. Construction of learning teams on the first day of class (step 11) randomly groups students of differing ability and interest and engages them in cooperative teaching and learning that potentially benefits each team member.

"Characteristics of the teacher" is probably the category most familiar to us as we design our courses, but at the same time, the most difficult to change. We should contemplate our familiarity with and attitude toward the subject—whether the subject is within or outside of our comfort zone—and our beliefs, values, strengths, and weaknesses as teachers and learners. All these characteristics vary widely among teachers, of course, but also can be different for a single educator depending on the course he or she is designing.

Lastly, Fink lists "the special pedagogical challenge." By this he means the special situation that challenges both students and teacher to create a meaningful and successful learning experience. An example for me would be the fact that my introductory science courses are primarily taken by students who are required to complete two science courses in order to graduate. Not only may they believe they have no interest in the subject, but also they may lack confidence in their ability to do well. This is where creative use of lecture-free teaching can help engage and reduce the anxiety of a reluctant student.

Finally, keep in mind that even if you teach several sections of the same course at the same school during the same semester, the situational factors can vary among those sections and necessitate differences in how you design and teach each section. We have all had instances where the morning section of a class runs smoothly and is enjoyable to teach, but the afternoon session leaves you feeling frustrated and incompetent. It is easy to blame this on a couple of less-than-cooperative students in the class, but paying attention to differences in other situational factors can lead to more successful and satisfying class meetings, saving you both time and emotional energy.

Step 2: Determine the Learning Goals for the Course

Too often as we plan a course, we begin by listing major content topics derived from the textbook's chapter titles: Our goal is to cover those topics. My current, student-centered approach goes well beyond having students learn a body of content. Instead my goals are more akin to those of a liberal education: I consider how to help my students significantly change their view of the world, how to foster an interest in the discipline that will continue beyond the date of the final exam, how to prepare students to make effective choices in the voting booth and become citizens of the world, and how to help them acquire thinking skills they will apply to other life endeavors.

Step 3: Create Formative and Summative Assessments That Provide Feedback About the Learning Goals to Both You and Your Students

Partway through a decade of pedagogical reform, I realized that my assessments needed to better reflect my learning goals. I originally anticipated completing the circle of my pedagogical reform by modifying my summative assessments, but after reviewing the extensive assessment literature, I realized I could not effectively use summative assessments without first introducing formative assessments, during which students practice the skills needed to achieve the learning goals for each course. Students come to class with prior knowledge and possible misconceptions that affect their interpretation of new knowledge. In-class formative assessments, usually ungraded and often completed within a cooperative learning team, can dispel misconceptions and contribute to new, more accurate learning.

These assessments provide information throughout the teaching and help students correct misunderstandings before they attempt a test or assignment on which they will be graded. This collaboration reinforces the "learning partnership" and is both a goal and a consequence of lecture-free teaching.

I suggest that formative assessments be a part of every class meeting and take a variety of forms. The best summative assessments require demonstration of skills and knowledge practiced during formative assessments by asking students to apply them to novel situations. Graded tests, papers, and group assignments should not only quantify student learning, but also enhance that learning. Occasionally, a student will excitedly tell me that while taking a test he or she thought about a concept in a different way and finally understood it. When that occurs I know that I have stumbled upon an effective evaluation of significant learning. I strive to have all of my assessments accomplish this.

Step 4: Choose a Teaching Strategy That Accomplishes the Learning Goals, Flexibly Responds to Feedback From the Assessments, and Maintains Coherence in the Course

When I began incorporating active learning into my teaching repertoire, I gave too little consideration to a teaching strategy. An early goal was to incorporate at least one active learning exercise into each of my lectures. The result was sometimes a disjointed class meeting comprised of activities that were relevant to content topics, but that did not always provide a smooth transition from one to the next. As I gained experience in lecture-free teaching, I created teaching strategies that supported the learning goals by encouraging students to prepare before each class meeting. One of my important responsibilities is to demonstrate connectedness among content topics and in-class activities.

Lecture-free does not mean that I never stand in front of the class and share factual information. I often give what could be described as minilectures, but these are in direct response to student questions or confusion rather than being short presentations I plan in advance to follow a prescribed schedule during a class meeting. I call this "student-stimulated teaching."

I provide students with outlines of content topics and important terms I expect them to understand. Having students take notes on these content outlines *before* each class effectively substitutes for my giving a lecture. Students' notes are derived from reading their textbook and ensure that they are informed of content I expect them to know, just as a lecture would. Although a student must be organized and disciplined to accomplish this reading and note-taking in advance, I experience greater success with this strategy than anticipating that my lecture will inspire students to do the reading afterward as promoted by others (Lord 2007). Reinforcing learning during class is not only a more efficient use of time but motivates students to attend class because they know I will not use class time to review what they could simply read on their own. My job is convincing students that preparing for and participating in class leads to better comprehension of concepts and ultimately saves them time.

Another consistent teaching strategy for all my courses is cooperative learning that begins in the first 15 minutes of the first class meeting of the semester. Many studies validate the effectiveness of cooperative learning (Lord 2001) and I have fully embraced this method.

Step 5: Develop In-Class and Homework Activities That Achieve the Learning Goals, Include Formative Assessments, and Support the Teaching Strategy

Years ago I decided that a goal for my classes was to develop my students' scientific reasoning skills, but I did yet appreciate that my courses displayed a disconnect between teaching strategies and learning goals. After a series of content-laden lectures, I created test questions requiring students to apply factual information to solve novel problems. My students, not surprisingly, could not make the leap and performed poorly on such questions. I realized that my instructor-centered strategy encouraged memorization of facts, while my test questions asked students to apply those facts in new ways. Today a task I particularly enjoy is designing activities that allow students to simultaneously learn and understand content.

As I stressed in step 4, activities should contribute to the cohesiveness of the curriculum. Equally important is avoiding too many different activities during the same class meeting. When first using active learning, I feared that activities would conclude too quickly, leaving me with extra time during which I would be forced to lecture or (Heaven forbid!) dismiss class early. Although I often have extra activities ready to go, I rarely use them, but instead reserve time at the end of the class meeting for closure and reflection on the activity's connection with previous and future topics. My change to fewer, longer class meetings each week permits me this important conclusion time because I do not spend time getting the class "warmed-up" on multiple days of the week.

I am alert for elements that tie the entire course together and with which I can reinforce earlier material while building upon the foundation. My goals are to connect all the course strands during class meetings near the end of the semester and to create appropriate comprehensive questions for the final exam.

Step 6: Decide on a Grading System for the Semester

To determine a course's grading system, I list all the tests, assignments, and other factors that will contribute to the semester grade. I then return to my previously determined learning goals (step 2) to be sure everything listed reflects those goals. To assign points to each item, I consider the comparative value I place on each facet of the course. The list of graded tests and assignments along with their point distribution are in the course syllabus. The relative number of points conveys to students the value I place on each category of assessment implying the time and effort they should invest. Too often students spend an excessive amount of time on something that does little to help them achieve the learning goals. I have no need to modify my grading system for penalties because the only official penalty I have is for late submission of work for which I subtract one point (usually from 30 or 35 possible points) for each day an assignment is late. I emphasize this policy both in the syllabus and verbally, thereby avoiding subjective decisions about giving extensions. Students rarely request extra time for completing a paper or argue with me about it. They make their own decision about whether it is worth it to turn in something late—another lesson in assuming responsibility.

Step 7: Assemble a Topic Schedule With Clearly Indicated Dates of In-Class and Laboratory Activities, Tests, and Due Dates for Homework Assignments

As students enter the classroom on the first day of the semester, I hand them a topic schedule (notice I do not call it a lecture schedule!). The topic schedule depends on elements I describe in steps 1–6, so is created after the completion of these steps. But since content outlines, syllabus, and coursepack organization depend upon the topic schedule, construction of this latter element must precede steps 8–10.

Step 8: Use the Course Textbook and Your Previous Lecture Notes to Create Content Outlines of Topics and Important Terms You Want Students to Know and Understand

Content outlines ensure that a course includes necessary and appropriate factual material and that students are informed about the sections of textbook chapters I consider most important. In other words, content outlines substitute for a lecture: Many of my content outlines are derived from past lecture notes, so they include exactly the same topics about which I would lecture. Student note-taking on outlines *before* class replaces student note-taking *during* class lectures. By guiding students through the chapters of the text, content outlines free class time for in-class activities that help students better comprehend and apply scientific information and uncover any misconceptions or confusion based on prior knowledge or from reading the textbook. On the first day of class, I verbally explain to students how to use the content outlines provided in their coursepacks. I also explain this in the syllabus. I format content outlines with four spaces between each term and concept to allow adequate room for note-taking as students read the assigned textbook chapter. Placing outlines with related in-class activities in a coursepack supports my teaching strategy goal of maintaining coherence (step 4).

Step 9: Compose a Detailed Course Syllabus Describing Features in Steps 1–8 and Emphasizing the Learning Partnership Among Students and You

As discussed in step 2, I did not truly understand the purpose and value of a well-designed syllabus until I read *The Course Syllabus: A Learning-Centered Approach* (O'Brien, Millis, and Cohen 2008). As these authors explain, a comprehensive, learning-centered syllabus describes not just what the instructor is going to cover but is an important resource to support learning and intellectual development. Composing such a document requires substantial thought and analysis; a syllabus evolves each time you teach a course to a different set of learners. The process of articulating learning goals, assessments, teaching strategies, activities, grading practices, and content help you develop and teach a better course.

Traditionally, faculty use the first day of class solely to distribute and review the syllabus. Before the semester begins, I alert students via e-mail that they will begin learning content on the first day of class and that this meeting will last for the full class period (almost three hours for my combined lecture/laboratory courses). Although I distribute both the topic schedule and the syllabus as students enter the classroom on that first day, I immediately engage students in an activity (step 11) that not only introduces them to some course content but also sends the message that attendance is important at all class meetings and that our time together will be used for learning that cannot occur outside of class.

Although I may discuss key sections of the syllabus, I mention several times that students' first assignment is to carefully read the syllabus. This is also listed as homework on the corresponding topic schedule. Those students who comply will find a surprise extra credit opportunity on the final page of the syllabus. At the beginning of the second class I announce how many students took advantage of this, thereby reinforcing the importance of students taking responsibility for their own learning.

My syllabus is generally six single-spaced pages (in addition to two pages of topic schedule) and all pages of this document are designed to be a resource for the entire semester. The syllabus not only provides practical information about the course, but also more general information (such as learning goals, teaching techniques, assignments, and grading system), always emphasizing the students' responsibility for their own learning and the ongoing partnership among the students and me.

Step 10: Organize the Topic Schedule, Syllabus, Content Outlines, and In-Class and Laboratory Activity Worksheets, as Well as Formative Assessments, in a Loose-Leaf Binder Entitled "The Coursepack That Students Bring to Each Class Meeting"

I have developed coursepacks for each course I teach and modify them as an important part of my preparation for an upcoming semester. Students buy coursepacks in the campus bookstore as one of their required texts. This arrangement not only saves my department money, eliminating the need to copy handouts, but I believe that purchasing even a modestly-priced coursepack increases students' commitment to the course. In addition, with all course materials in their possession from the first week of the semester, students can plan ahead and know exactly what they missed if absent.

Most importantly, with a coursepack, students can view the structure of the entire semester. Demonstrating coherence is a critical part of my teaching strategy (step 4) and over the years I have become more skilled at choosing appropriate activities so that individual class meetings not only have a theme and make smooth transitions between activities, but also ensure that students experience connections among the content topics. Having all the semester's course elements in chronological order in a single loose-leaf notebook underscores these links.

Step 11: Construct, on the First Day of Class, Heterogeneous Learning Teams of Four to Five students

The first day of class is, in many ways, the most important class meeting of the semester. Investing time and effort to plan an interesting, well-organized day can create a positive classroom climate that lasts the entire semester. The first class is your opportunity to set the stage for constructive interpersonal relations among classmates and between you and your students.

Goals for my first class meeting include having students: (1) understand the benefits of my lecture-free teaching methods; (2) experience how the course is structured and what is expected of them; (3) learn some course content; (4) begin to feel comfortable among their classmates and to develop a rapport with me; and (5) understand their responsibility for the success of the course. My method for building heterogeneous learning teams initiates work toward all these goals within the first 15 minutes of the semester. For each course, I design a different organizational concept by which students form cooperative learning teams of four or five students with whom they will work on in-class and homework activities for the semester (Wood 2007; Wood 2009a; Wood 2009b).

Because of previous negative experiences, some students are reluctant to work in groups. Explaining to them, at the outset, the proven benefits of teamwork and also how I will monitor, evaluate, and assess contributions of team members starts them off with a more positive attitude about what may be an unfamiliar or uncomfortable learning tool. The theme of the team-building activity introduces factual content that is reiterated throughout the semester, so this exercise is much more than an icebreaker.

I believe in the power of words and that your choice of words affects listeners' or readers' perceptions and expectations of what you are trying to communicate. For years I talked to students about forming "cooperative learning groups," an important instructional strategy first promoted in the 1980s (Johnson and Johnson 1989; Johnson, Johnson, and Smith 1991). Now I prefer the expression "learning teams" (Michaelsen, Knight, and Fink 2004). This term is meaningful to my often sports-minded students and describes a supportive interaction among students that ultimately benefits everyone.

Step 12: Use Class Time to Answer Student-Generated Questions and to Lead Activities and Laboratory Exercises That Contribute to Comprehension of Concepts

I invite students to promptly communicate with me about course content they do not fully understand. There are several different ways to accomplish this: (1) e-mail me questions before 7:00 a.m. of the class day; (2) write "the murkiest point" at the end of a class meeting; (3) call or visit me in my office; and (4) use the time-honored method of raising one's hand during class. The first thing I do at a class meeting is respond to "murkies" from the previous class, as well as to any questions I received via e-mail. In the syllabus I explain that if I receive no questions or requests for additional explanation of a topic, I assume everyone in the class understands both the previous and new material. I then proceed with the scheduled in-class activity or laboratory exercise. As I describe in step 4, I sometimes give what could be described as minilectures, but these are in direct response to student questions or confusion.

During a semester, I employ a variety of instructional strategies: inquiry-based laboratories resulting in scientific reports; writing by students; cooperative learning including peer-instruction, problem-based learning, and test review sessions; team-based learning; service learning; case studies; and student-led teaching models. As discussed in step 5, a plethora of classroom-tested activities for science students at all levels is available in journals, books, and online sources. I keep in mind that what I have the students do during class should increase their understanding and give them practice applying factual knowledge related to the day's topic. Although with lecture-free teaching, one cannot predict exactly how long a planned exercise will take, I reserve time for clarification and closure at the end of every class.

Class time should be used for learning that students cannot accomplish alone outside of the classroom. Students in Light's (2001) study reported that classes where a professor simply repeats what they have just read or could easily read in a textbook is not a good use of their time. Similarly, students will not take in-class activities, laboratory exercises, videos, and guest lectures seriously unless test questions and homework assignments reflect these experiences (Tobias 1990).

Step 13: Enjoy the Unpredictable! Keep Your Mind and Eyes Open for New Teaching Techniques and Activities That Augment the Cohesiveness of Your Course, Support Your Learning Goals, and Stimulate a Dialog Between You and Your Students

Although the unpredictability of leading a lecture-free class may at first be unsettling, I encourage educators to embrace the energy that accompanies this pedagogy. Flexibility and responsiveness to what happens during each class meeting with each unique group of students is the key to the success of my methods.

I began my teaching reform by gradually incorporating the ideas of others about which I read or heard. As I gain experience and confidence, I use more of my own ideas and am sometimes surprised by my creativity. An everyday experience or conversation may suggest an ideal in-class activity or laboratory investigation. I am increasingly bold about trying something completely new, but afterward ask students to reflect on what we have done and tell me whether it was helpful to them, if I should repeat it with future students, and their recommendations for improving the activity.

Honest feedback from those who are recipients of my teaching methods is vital. We ask our students to take risks in their learning, and we must do the same. My hope is that as we develop rapport and mutual respect, students and I feel more like equal participants in the process and that we truly engage in a learning partnership.

What Lecture-Free Teaching Can Accomplish for Science

The United States today has an urgent need for science educators and researchers to replace retiring professionals and to meet the requirements of a society increasingly more dependent on science and technology. Equally important is the need for a more scientifically literate citizenry able to function in a technologically complex world, not only by understanding scientific concepts and new discoveries, but also by making informed decisions in their personal lives and in the voting booth. I believe that fundamental reform of science teaching methods can simultaneously address these issues.

The scientific community benefited during the post-Sputnik period with an increase in both material resources and citizen support. But concomitant reforms of science education consisted mainly of altering course materials and adding teaching enhancements: These are not factors that contribute to students deciding for or against science careers.

We need real reform in science education and we need it now. The reform must grow out of collaboration among educators at all levels. The educators, in turn, need to listen to what their students say about classroom methods that work for them. If we truly want to increase numbers of science majors and create a more scientifically literate society, we must tap resources that have in the past been ignored: We need to engage not only the type of students who have always been successful with traditional lecture and note-taking, but also those who respond better to different modes of teaching. We need to include Tobias's "second tier" (1990) in an expanded potential talent pool by responding to considerable data recommending that we replace lectures with more interaction among students; memorization with conceptual understanding; and cookbook- style laboratories with inquiry-based exercises. Such recommendations are based on scientific evidence, and when educators respond to this evidence, they are "teaching science scientifically" (Dehaan 2005). I encourage college science educators to share their innovative teaching methods in journals and books such as this so that we can all contribute to our common goal of improved science education for all.

References

Dehaan, R. L. 2005. The impending revolution in undergraduate science education. *Journal of Science Education and Technology* 14 (2): 253–269.

Fink, L. D. 2003. *Creating significant learning experiences: An integrated approach to designing college courses.* San Francisco: Jossey-Bass.

Grunert, J. 1997. *The course syllabus: A learning-centered approach.* Bolton, MA: Anker Publishing.

Handelsman, J., D. Ebert-May, R. Beichner, P. Bruns, A Chang, R. DeHaan, J. Gentile, et al. 2004. Scientific teaching. *Science* 304 (5670): 521–522.

Johnson, D. W., and R. T. Johnson. 1989. *Cooperation and competition: Theory and research*. Edina, MN: Interaction Book Company.

Johnson, D. W., and K. A. Smith. 1991. *Active learning: Cooperation in the college classroom*. Edina, MN: Interaction Book Company.

Light, R. J. 2001. *Making the most of college: Students speak their minds*. Cambridge, MA: Harvard University Press.

Lord, T. R. 2001. 101 reasons for using cooperative learning in biology teaching. *The American Biology Teacher* 63 (1): 30–38.

Lord, T. R. 2007. Teach for understanding before the details get in the way. *Journal of College Science Teaching* 36 (6): 70–72.

Mazur, E. 2009. Farewell, lecture? *Science* 323 (5910): 50–51.

McManus, D. A. 2005. *Leaving the lectern: Cooperative learning and the critical first days of students working in groups.* Bolton, MA: Anker Publishing Company.

Michaelsen, L. K., A. B. Knight, and L. D. Fink, eds. 2004. *Team-based learning: A transformative use of small groups in college teaching*. Sterling, VA: Stylus.

National Research Council (NRC). 1996. *National science education standards.* Washington, DC: National Academies Press.

O'Brien, J. G., B. J. Millis, and M. W. Cohen. 2008. *The course syllabus: A learning-centered approach*. 2nd ed. San Francisco: Jossey-Bass.

Tobias, S. 1990. *They're not dumb, they're different: Stalking the second tier.* Tucson, AZ: Research Corporation.

Wiggins, G., and J. McTighe. 2006. *Understanding by design*. 2nd ed. Upper Saddle River, NJ: Pearson Education.

Wood, B. S. 2007. Learning biology while constructing cooperative learning groups: The first 15 minutes. *The American Biology Teacher* 69 (6): 330.

Wood, B. S. 2009a. Learning science while constructing learning teams. *Journal of College Science Teaching* 38 (5): 28–32.

Wood, B. S. 2009b. *Lecture-free teaching: A learning partnership between science educators and their students.* Arlington, VA: NSTA Press.

Requiring College Students in a Plant Science Course to Take Control of Their Learning

Thomas R. Lord
Indiana University of Pennsylvania

Setting

The plant biology course discussed in this chapter was designed for science majors, most of whom were biology and secondary biology education students. The course was taught by faculty of the biology department at Indiana University of Pennsylvania and follows the recent recommendations put forth by the American Association for the Advancement of Science, the National Research Council, and the National Science Teachers Association. Instruction in the course is designed around the constructivist teaching model based on inquiry instruction.

Constructivism is not new to the author. He has published over two dozen articles on the subject and presented workshops on inquiry teaching at conferences of the National Biology Teachers Association, National Science Teachers Association, American Association for the Advancement of Science, and the NSF Chautauqua Short courses.

Higher Education

This is an amazing time to be in higher education. College science instruction is undergoing changes that are long overdue. Several months ago, the American Association for the Advancement of Science (AAAS), supported by National Science Foundation (NSF) and the National Research Council (NRC), issued a document that outlines the pathway that college science instruction should be directed. *Vision and Change in Undergraduate Biology Education: A Call for Action* (Brewer and Smith) examines the factors that have created this need for change over the last few decades. The document points out that scholarly information has greatly increased; that the use of graphics in learning materials has shifted from black, white, and gray to multitudes of colors; and photos, charts, and graphs have become more realistic in their design. The authors also state that student populations have changed over the past 20 years and, along with students, the greatest changes during the last two decades have occurred in the various forms of technology that have directly affected teaching. Technological advances have revolutionized the way knowledge is gained and the way we learn. It is no longer necessary, for example, for students to spend hours in their dorms and libraries memorizing facts, terms, and definitions. Such information is now just a click away on a multitude of tiny handheld devices.

Not only is scientific nomenclature at our fingertips but the instructions for using the information is also very easily accessed. This past weekend, for example, I wanted to find information about *Listeria*, a bacterium that is commonly found in soil and water that is sickening people around the world. A question sent to various search engines on my computer quickly revealed several articles written by scholars about the epidemic, explaining its source, defining its incubation and describing the way it affects people. Information is so easily accessed that students have pretty much have stopped trying to learn the information. Because of this, students in class are less attentive and more distracted by secondary off-topic information.

This has lead educational leaders to recommend that teachers and professors bring students into the knowledge-gaining process as active participants. In both the *Next Generation Science Standards*, developed for precollege pupils, and the *Vision and Change* document, designed for postsecondary science students, contemporary instructors are encouraged to involve their students in hands-on, minds-on interactions that inspire them to actively figure out the answer rather than to passively copy it down from the instructor's lectures. Research has found that actively challenging students to discover answers leads to large information gain (Handelson, Miller, and Pfund 2008; Ebert-May and Hodder 2008; Hoffer 2009).

Experiential Learning

Active student learning also allows the teacher to appreciate how new information is applied. In my science class, for example, I no longer lecture about the fact that adrenalin is released from nerve endings and the adrenal glands in people who are undergoing stress. Instead, to demonstrate the point, I tip over a cardboard box that I've brought with me to class to release a large snake onto the class room floor. This action immediately creates stress in the class. After the commotion dies down and I've returned the reptile to the box, I asked the students to write down 10 responses they felt in the first 15 seconds upon seeing the snake. Students easily write down their reactions, which coincide physiologically with the "flight or fight" reaction, an autonomic response for the human peripheral nervous system. More importantly, the students tend not to forget that experience. I've had students tell me they remember that event for years and that the hormone, adrenalin from nerve endings and the adrenal gland, is responsible for their reaction. Before I used this tactic, most students in my class could not remember this information the day after I had mentioned it in a lecture. As stated earlier, research shows that college science instructors need to increase the level of awareness of the students in their classes to ensure that lasting learning takes place.

The same concept of active learning holds true in my biology laboratory. For years I had the students follow printed directions from commercially prepared laboratory books. The students would collect lab glassware and instruments printed in the materials section of the exercise and follow serial instructions on the day's experiment, periodically jotting an answer on the space provided in the lab directions. At the conclusion of the experiment, students would turn in their completed lab report, which most of the time was the direction sheet that they had carefully torn from their commercially prepared, and usually expensive, lab book. Students taught through this serial response method rarely interpret their data with charts and graphs unless the lab directions had told them to do so.

Using Contemporary Technology With Experiential Learning

Experiential changes encouraged by the nation's leaders are designed to bring the students into the investigative procedure. No longer are students to passively follow cookbook directions from commercial lab manuals. Instead, following recommendations of the 21st-century learning skills (Figure 3.1), professors require students to research information and present it to their classmates using sophisticated teaching technologies like graphic whiteboards and 3-D interactive visualization software, which allow students to manipulate images, and to meet students virtually within or outside the institution with collaborative technologies like Blackboard, Course-web, or Web-interactions. In my classes each student researches a course component and presents it to their classmates with various presentation and interactive software (Van Gundy et al. 2009). For example, when the topic of fluid transport in plants is taught, a member of the class, rather than the professor, leads a discussion of vascular anatomies and fluid dynamics that they have researched (and often discussed with the professor).

Figure 3.1. Supporting Components for Instructing 21st-Century College Science Classes

Old Paradigm

- Acquiring information and recitation of acquired knowledge
- Using summative tests of discrete, factual information that is easily measured
- Assessing to learn what students do not know, and assessing only achievement
- Rigidly following curriculum
- Presenting knowledge with lecture, text, and demonstration
- Asking for recitation of acquired knowledge
- Providing textbook and lecture-driven curriculum with broad coverage of unconnected factual information.
- Seeing teachers as based in classrooms, learning alone
- Separating theory and practice
- Transmitting teaching and content knowledge through lectures and reading
- Seeing teachers as consumers of knowledge
- Providing one-shot sessions, courses and workshops to teachers as technicians

New Paradigm

- Treating students alike and responding to them as a whole
- Maintaining responsibility and authority by the teacher and supporting competition rather than collaboration
- Learning opportunities that favor one group

- Understanding scientific concepts, developing abilities of inquiry, and learning subject matter disciplines in the context of inquiry, technology, science in personal and social perspectives, and history and nature of science
- Assessing to learn what students do understand, as well as achievement and opportunity to learn
- Selecting and adapting curriculum
- Guiding students in active, extended scientific inquiry
- Providing opportunities for scientific discussion and debate among students
- Providing curriculum that supports the standards, includes a variety of components (e.g., laboratories, emphasizing inquiry and field trips) and includes natural phenomena and science-related social issues that students encounter in everyday life
- Treating teachers as professionals and as members of collegial communities
- Integrating theory and practice in the school setting
- Encouraging teachers to learn about science and science teaching through inquiry and investigation
- Employing long-term coherent plans including a variety of activities for reflective practitioners
- Seeing teachers as producers of knowledge
- Providing opportunities both for continual learning and networking for school improvement

Support Structures for Professors

- Brainstorming and concept mapping software
- Online authoring, brainstorming, graphics, spreadsheet and presentation software
- Online collaboration, conferencing, communication tools
- Resources in the local community including people, places, institutions, and information
- Social networking sites
- Media creation tools including software for graphic design, digital photo and video editing, and presentations.
- Online collaboration, conferencing, communication tools for authentic research collection, as well as online data sets with peers and scientists
- Use of probeware, mobile media devices, GIS and various online tools for data
- Social networking sites
- Electronic portfolios
- Online conferencing, communication tools
- Social networking sites
- Media creation tools including software for graphic design, digital photo and video editing, and presentations
- Access to the web and personal computing
- Brainstorming, concept mapping software
- Computer-aided design, modeling software, and simulation software
- Digital production tools (digital photography and video) ,GIS and GPS tools
- Graphics software (drawing, painting, image editing); interactive online sites, multimedia resources (clip art, video)

- Ongoing professional development to promote an inquiry approach in the context of laboratory and field, as well as through use of technology
- Collaboration, conferencing, communication tools (online)
- Social networking tools
- Online courses and self-paced learning modules

Active Learning in the Laboratory

As mentioned earlier, the methodology for teaching the laboratory component of the course involves the students in a different way as well. Following an inquiry approach to lab experimentation, professors often pose an investigative question that students need to answer during the lab period. Gathering around a square lab table, three or four class members discuss ways they can answer the challenge during the lab session. For example, students could be challenged to investigate if common bread mold *(Rhizopus stolonifer)* prefers to grow on specific kinds of bread. After students have discussed the challenge with their teammates, they formulate a hypothesis and devise their plan. With their investigative scheme in mind, the group members gather the materials required to do their experiment from a stock closet in the room (day-old breads are easily obtained free from food markets), set up the experiment, and collect their data. Teams often approach the challenge in different ways that may or may not direct them to the correct answer. In this lab, for example, some teams may not moisten the bread, control the temperature, or provide a suitable container for the incubation process, which will certainly hamper the results. Therefore, during this investigation, the constructivist professor would move about the class, asking questions or making comments that will redirect groups that are severely off the mark. Alternatively, the professor will encourage members from groups that have successfully completed the task to move to teams that are having trouble carrying out the challenge. The accomplished students are directed not to tell the answers to the troubled group but rather guide them with questions and suggestions to reach the successful completion. This student-teaching-student approach has been shown to effectively increase understanding and is strongly encouraged in the new standards (Brewer and Smith 2009).

At the conclusion of the investigation the experimenters observe their results and draw their conclusions. In addition to writing up their efforts, students are also required to chart or diagram their results. The findings of the student group will enable them to either support or disprove their hypothesis.

Experiential Labs

At the onset of the course, performing a lab investigation without cookbook directions has been shown to be stressful for the participants. As students become familiar with experimental design, they gain confidence in their ability to use investigations as a means of understanding and learning the discipline. Within weeks of starting the course, student attitudes change and by midterm they tend to favor the approach of creating their own method to solve the question over the tiresome cookbook method. As they do this, team members build an understanding of the nature of science, an objective highly supported by the new Standards and the 21st century learning skills (Van Gundy et al. 2009)

As an example, students in my plant science lab were asked to find if seeds of closely related species (i.e., lima beans, kidney beans, navy beans, black beans) germinated at the same rate. Each team of students ($N = 3$ to 4) immediately began to discuss how the experiment should be performed to answer the question and generate a hypothesis on the outcome of the investigation. The team also considered the materials they needed to maintain their investigation and began to set up their experiment. As they did this, the participants noticed variables that would influence the outcome and devise ways to control them. Once they had established the means of running the investigation, the experimenters gave their setup a test run, and if it functioned as planned, the students began to collect their data (if the experimental routine did not work, they adjusted the procedure until it did). At the end of the experiment, the participants drew (discussed) a conclusion, diagrammed their results, and verified or cast out their hypothesis.

Reporting the Results From Experiential Labs

The reporting of the learning activity is also done in nontraditional ways. Instead of simply filling in the student's observation on a line in a commercially prepared laboratory manual, students describe their investigative setup and record their data in a nonscripted lab book or, if outdoors, a nonscripted field notebook. The reports are descriptive about what the students did and suggest what statistical treatment the team should apply for correct conclusions to be drawn. The point is reached when the lab notes are converted to reports that not only the professor evaluates, but several times a semester, student peers will read as well. The peer readers are equally capable participants who are in a different laboratory section of the class or, in some instances, a completely different science course (i.e., lab write-ups of plant science students may be exchanged with students in plant ecology) and who have also designed their report as a descriptive dialogue.

The reports describe the initial investigative design of each group in their attempt to answer the challenge question posed by the course instructor. Teams also defend their hypothesis and their procedure, as well as how they derived their outcomes. Results are always diagrammed or plotted on charts, and conclusions are justified and supported with current literature from reputable sources. Within the week, reports are returned to the initial investigators who are expected to answer the reviewer's comments through defense of the team's conclusions. Such peer review of research mirrors that done in higher level, contemporary science and is strongly advocated in the *Next Generation Science Standards* and the *Vision and Change* document.

I have tried an interesting alternative to a written description of the experiment. Investigators are told to capture the planning, setup, and experimental procedure of the student-designed experiment as a video recording rather than in a written composition. All required aspects of the written report still need to be followed in the video; thus group discussion, formulation of a hypothesis, and collection of materials necessary to perform the experiment should be recorded. The video also exposes the investigative procedures followed by the experimenters and how the teams formulated their discussion and determined their conclusions. The videos were not only graded on experimental design but also on quality of the production.

Encouraging Dissemination of Their Findings

The investigative teams are encouraged to submit journal-quality reports for further scrutiny through presentation at professional meetings or submission to respected scientific publications. Most areas of the country support state or regional science meetings or fairs where student research initiatives are shared with others. Various regions of the nation, for example, hold science teacher conferences and some organizations, such as the state chapter of the Academy of Science, hosts student research competitions for secondary and postsecondary researchers. National-level science and engineering organizations, also hold annual conferences, most of which have minimum registration fees for student attendees and provide opportunities to report what they, the students, have discovered with their research.

In a similar fashion, over the years several students shared their video with attendees at a science conference or meeting. Despite the fact that there isn't typically a category for the entries, conference officials generally allow teams to enter their video in the conference's poster competition. The experimenters generally display their video on a small, black-and-white television with a built-in video playback unit. Despite the marginally acceptable quality of the tapes, viewers at the exhibit enjoy witnessing the student's preoperational planning of the procedure and applaud the team's innovative entry. Unfortunately, judges of the competition are generally not able to agree on the criteria to evaluate the effort and the videos usually do not receive an award.

Measuring the Outcome of the Course

The basis for the initiative was grounded in a document written a decade ago by researchers at the American Society of Plant Biologists. Entitled *Principles of Plant Biology* (1999), authors Mintzes, Wandersee, and Novak developed a dozen principles that all courses in plant biology should strive to meet. Embedded in the text of this document is the Measure of Plant Science Knowledge, a quantitative instrument designed to provide a teacher or professor with a means of assessing the gain in understanding in students taking a plant science course. Designed by James Wandersee (2000), the instrument is constructed as equated pretest and posttest questionnaires. When used in conjunction with compatible control and experimental populations, significant gains in understanding with the intervention are revealed.

The content design of this study uses modified versions of the Wandersee instrument to assess factual knowledge (Figures 3.2 [p. 40] and 3.3 [p. 41]). Questions on the exams varied from recall (i.e., *Where on the lifecycle diagram does meiosis take place?*), to requiring students to apply information to situations they had not seen before (i.e.: *What is described by this diagram and how would you explain the sudden rise of line B in the center of the diagram?*). When the resulting scores on the pretest (version 1) were compared between the control and the experimental groups, no significant difference between the two populations were found. Students taught this semester with traditional methods answered as well or better than the students taught with the inquiry intervention on the initial pretest questionnaire.

Answers on the questions on the posttest (version 2), however, differed significantly between control and experimental student groups. Students taught with hands-on, minds-on instructional techniques performed *significantly* better on the questions when compared to students taught through lecture/recitation. Most students, for example, had no trouble applying information from a chart or graph to a natural event (i.e., *How do you account for the high stoma activity in the middle of the night, as indicated on the graph*?) or using what they learned in class to questions in the field (i.e., *How would you explain the different growth patterns seen in the plants growing on opposite sides of the street*?). Mean scores on the questions of the inquiry students were quite a bit higher than the scores from traditionally taught students (Table 3.1, p. 42).

Figure 3.2. Preinstructional Test (* denotes an additional question not found in the original document)

CODE:____________

Directions: Write the term or phrase that completes the statement or answers the question.

1. More than 500,000 different species of plants have been identified. How are plants identified?
2. Roots are vital to a plant; the survival of a plants' roots depend upon ...
3. Give 5 factors that are involved in plant respiration.
4. A flower has a bright yellow organ located in its center (this plant structure is called a stigma) On what basis would you describe its function?
5. Water performs many vital functions in plants; list 5 of them.
6. Plants play an important role in the support of all living matter, they are a dominant force in our ecosystem partially due to ...
7. Why do seedlings of some indoor plants grow toward the window?
8. Are reproductive organs in some plants both sexes?
9. What is the concept a biologist would use to explain the reason for a plant wilting?
10. Plants have specific needs that are essential for plants to survive. List 5 of them.
11. Distinguish between seeds from a monocot and a dicot.
12. In a food chain involving green plants, insects, birds, and mammals, the original source of energy is ...
13. What is the connection between trees that form root nodules and trees that do not possess nodules?
14. A chimpanzee eats bamboo shoots, fruits, and other plants in the jungle. How would a biologist describe how the energy that flows through the biosphere affect the animal?
15. The green layer of algae that is often found floating on the tops of ponds has several functions. List 3 of them.
16. Give 3 functions that describe the role of fruits in the plant cycle.

17. The amount of sunlight that reaches the floor of a tropical rain forest is considerably lower than that of a grassland. How does this difference affect the organism found in the areas?
18. What are limiting factors in the survival of plant species?
19. A plant is placed in a bottle with only tiny air holes at the top. The plant is not watered, yet droplets of moisture are observed on the inside of the bottle, this is due to …?
20. What are 5 factors that influence the size that a tree could grow.

*21. Where on the lifecycle diagram does meiosis take place?

*22. What is described by this diagram and how would you explain the sudden rise of line B in the center of the diagram?

Figure 3.3. Postinstructional Test (*denotes an additional question not found in the original document)

Directions: Write the term or phrase that best answers the statement or answers the question.

1. A biologist discovers this flower on a plant hunt. What factors could he use to identify it?
2. Plants receive nutrients through the process of…?
3. Cellular respiration is a series of chemical reactions that break down organic materials and release energy. If the temperature increases, what would we expect?
4. The flower has brown structures in the center. This structure is a part of the stamen that produces pollen; what is it?
5. The main tissue involved in transporting water from roots to leaves is the…?
6. Suppose the intensity of sunlight was drastically reduced for several months due to volcanic ash from an erupting volcano. How could the following members of the ecosystem be affected:grasses, rabbits, and hawks?
7. The plant that is growing toward light is demonstrating a plant's response to the light source called …?
8. In flowering plants, the process that enables the sperm to approach the egg is …?
9. List 3 well-known pairings of plant structures/functions.
10. Plant survival depends on several things. List 5 of them.
11. At what point in its life cycle does the plant begin to manufacture its own food?
12. Climate change will have the least effects of what 3 items in plants?
13. Monocots and dicots have several things in common. List 3 of them.

14. Plants and animals have a symbiotic (interdependent) relationship with respect to ...?
15. A bromeliad (flower) that grows from the side of a tree is an example of what type of relationship?
16. What happens to the oxygen that is involved in photosynthesis?
17. Photosynthesis includes several different processes. List 5 of them in order.
18. Roughly two-thirds of all vascular plants are found in the tropics. List 5 factors that do not affect their abundance.
19. A plant is placed in a dark room, the stem of the plant becomes weak and the plant begins to fall to the side, the plant is measured and has not grown since its placement in the dark room. How would a biologist explain the condition of the plant?
20. "Plants are the center of our existence." List 3 factors that support the statement.
21. How do you account for the high stoma activity in the middle of the night, as indicated on the graph?
22. How would you explain the different growth patterns seen in plants growing on different sides of the street?

Table 3.1. ANOVA on Mean Scores on Plant Questionnaire Between Control and Experimental Groups

Source	SS	MS	F	p
Between groups	16.5	8.15	4.11	<0.05
Within groups	29.7	1.98		

Correlation Between the Pretest and Posttest Scores on the Plant Questionnaire (ns=20)

Group	Covariance	R^2	r	p
Control	1.47	0.51	0.714	ns
Experimental	0.66	0.22	0.468	<0.01

Measuring Outcomes in the Field

Students in both groups were also evaluated on their practical knowledge of plant science through a field test consisting of 50 questions. Students in the study were asked to write a short answer concerning an environmental occurrence on a 3" × 3" paper card that was part of a 50-page stack designed as the test booklet. After a participant had written his or her answer on the card, it was carefully ripped from the stack and handed to a teaching assistant who clipped it to a bundle of other students' answers and slid it into a cloth bag. A typical question on this practical would be: *Provide three possibilities for the creation of the large curve midway up the trunk of this young maple tree,* or *How old is this tree branch and what could have caused the large scar midway up the branch's shaft?* The test was designed to assess whether the students could use information they had learned in the course in an outdoor setting.

The students in the inquiry-directed population did well on the questions on the field exams. Although the participants had never been to the test site, they were able to apply the content they had learned during the class to environmental situations they had never before seen. The traditionally taught students, on the other hand, had trouble answering many of the practical questions. They generally knew the name of the plants in the question asked during the exam but had trouble applying the answer to a situation asked by the instructor. Another good example of the difference between the two groups of students occurred on a question involving a red maple and tulip poplar trees. The two were growing closely next to one another with a section of their trunks united a foot off the ground. The union continued for at least two feet before the trunks parted and continued separate growths. The instructor asked the students to *(1) identify the trees, (2) give at least two scenarios as to how the union of the trunks could have occurred and (3) what the internal physiology/genetics within the merged section would be.* The inquiry-taught students had only modest problems answering the questions and did well overall, while the traditionally taught participants could only identify the tree types and wildly guessed about the second and third parts of the question (Table 3.2). Several students mentioned after the exam that they thought the exam was terribly unfair since they had not previously been in the location where the test was given.

Table 3.2. "t" Ratios Comparing the Scores on the Field Final Between the Experimental and Nonexperimental Groups

Comparison	M	SD	t	p
Experimental and Control	2.0	0.6	1.93	<0.05

Assessing Attitudes

To evaluate student's attitudes about the plant science course, the members in the class were given a free-response survey to compare different aspects of the class to their experiences in previous science courses (Figure 3.4, p. 44). Most of the students enjoyed the laboratory and field experience, and liked working with teammates rather than alone. A good number of the population who used "cookbook" lab exercises felt betrayed by questions on the tests that the class had not discussed before. It was apparent that most of the students in the control group believed that only course content discussed by the professor in class should appear on an exam.

The results were very different with the members in the inquiry-taught section. Most of the students enjoyed the challenge of using the information they had learned and were disappointed with recall questions. One student wrote how surprised she was when she could answer a question about an insect's niche by noting structural features it possessed. It was clear that the inquiry-taught group understood the nature of science and could apply what they had been taught (Table 3.3, p. 44).

Figure 3.4. Student Survey for Evaluating Differences Between Didactic and Inquiry Teaching

(Answer yes or no followed by one or two short sentences to explain your answer.)

1. Do you think you learn more information working in teams or by yourself?
2. After being given the lab or field objectives, would you rather do your own investigation procedure or work with teammates?
3. After being given the lab or field objectives, would you rather create your own system to accomplish them or do you prefer following step-by-step directions?
4. Do you believe you understand the nature of science after designing your own lab?
5. Do you believe you understand the scientific method after designing your own lab?
6. Do you feel you understood the lab more when you designed it yourself?
7. Did you enjoy writing your own labs rather than being given instructions?
8. Do you think you can recall information for a longer period of time when you design your own lab/field activities?
9. Do you think it is more beneficial for the instructor to point out characteristics of plants or for you to discover them with your teams?
10. Do you like learning that emphasizes understanding rather than memorization?
11. Do you like learning in a practical way or in a content-recall way?
12. Did you like the noncompetitive method of teaching or would you prefer a more competitive method?
13. Did you find the lab preview the instructor gave to be beneficial?
14. Did you find the lab and the field reports beneficial?
15. Do you like having quizzes each session or would you rather have larger tests?
16. Did you feel the student presentations helped you to learn the information?
17. Did you like having refreshments during the presentations?
18. Do you like learning in the field better than learning in a class/lab room?

Table 3.3. "t" Statistic Comparing the Means on the Plant Attitude Survey on Pretest and Posttest Scores of the Experimental and Nonexperimental Groups

Group	Mean	"t" Value	p
Control	0.25	1.56	ns
Experimental	1.60	8.72	<0.01

Conclusions

Overall, the study clearly brings out one of the major flaws in higher education: Students believe they have *learned* when they can recite on an exam the information presented to them in class. Researchers have noted, however, that *real* learning is not witnessed in students who can't take what they think they know to the next level. Quality education provides a person with the ability to *use* the content he or she has been taught. Recognizing this, it's clear that many of nation's schools, colleges, and universities have a long way to go before they provide students with a functional education.

References

Brewer, C. A., and D. Smith. 2009. *Vision and change in undergraduate biology education: A call for action*. Washington, DC: American Association for the Advancement of Science.

Ebert-May, D., and J. Hodder, eds. 2008. *Pathways to scientific teaching*. Sunderland, CT: Sinauer Associates.

Handelson, J., S. Miller, and C. Pfund. 2008. *Scientific teaching,* New York: Freeman and Company.

Hoffer, W. 2009. *Science as thinking: The constants and variables of inquiry teaching*. Portsmouth, NH: Greenwood Publishing Group.

Mintzes, J., J. Wandersee, and J. Novak. 1999. Principles of plant biology. *http://my.aspb.org/?page=EF_Principles&CFID=860774&CFTOKEN=32635626*

National Research Council (NRC). 2012. Next generation science standards for today's students and tomorrow's workforce. In *Framework for K–12 science education*. Washington, DC: National Academies Press.

Van Gundy, S., A. Hovey, S. Koba, S. Lee, and L. Mayo. 2009. 21st Century Skills Map Task Force.

Wandersee, J. 2000. *Designing an image-based plant science test: (V1 and 2)*. In *Assessing science understanding: A human constructivist view*, ed. J. J. Mintzes, J. Wandersee, and J. D. Novak, p. 137. San Diego, CA: Academic Press.

Peer-Led Study Groups as Learning Communities in the Natural Sciences

Claire Sandler
University of Michigan

Joe Salvatore
University of Michigan

Setting

The University of Michigan in Ann Arbor (UM) is a comprehensive graduate degree–granting public institution located in a midwestern college town with a population in excess of 114,000. The UM is a highly selective university as evidenced by the composition of the fall 2011 entering class, whose median GPA was 3.90, median SAT math and verbal total was 1350, and median composite ACT was 30. The UM is a national public university with a student body largely drawn from the state of Michigan. The UM student body also features a sizeable number of students from the surrounding states of Illinois, Indiana, and Ohio, as well as a significant number who call the states of New York, California, Florida, and Texas home. In fact, UM students represent all 50 states as well as nearly 100 foreign countries. The largest enrollment of undergraduates at the UM is found in the College of Literature, Science, and the Arts (LSA) with approximately 19,000 students of the campus total of over 27,000 and 1,100 faculty members. LSA's fall 2011 entering class enrolled 4,412 students.

Learning Community Context

In 1998, the Boyer Commission on Educating Undergraduates in the Research University issued its report *Reinventing Undergraduate Education: A Blueprint for America's Research Universities* in which it identified 10 basic reforms that research universities needed to undertake in order to fulfill their obligations to undergraduates. One of the 10 tenets upon which its recommendations were built was that "a sense of community is an essential element in providing students a strong undergraduate education." The Peer-Led Study Group (PLSG) program detailed here helps to provide a framework upon which a sense of community is built for the thousands of students at the UM who are enrolled each term in introductory science courses.

Over the past 20 years, it has also become a widely accepted notion that undergraduate learning is enhanced when students engage in both academic and nonacademic activities, both classroom-based as well as outside the classroom (Astin 1993; Pascarella and Terenzini 1991).

More recently, additional evidence has been compiled that further supports the idea that to promote student learning institutions ought to provide opportunities for students to spend more time engaging in academically productive activities outside of the classroom (Kuh et al. 2005). It is also notable that the importance for learners to be *actively engaged* in learning has been cited as a "basic tenet of modern cognitive theory" by Barkley, Cross, and Major (2005).

The UM's College of LSA created the Science Learning Center (SLC) in 1989 to provide structure for outside-of-classroom academic programming and support for students enrolled in introductory natural science courses in the context of a nurturing community of learners.

Today, the SLC's support for undergraduate science learners takes a variety of forms including: providing access to instructional computing labs, study lounges, and drop-in help sessions; sponsoring science study skills workshops; operating a lending library of learning tools and instructional materials; and operating the PLSG program. So, at its core, the SLC provides a place where students can spend time before and after classes engaging in unstructured, but academically oriented activities. With ample individual and small-group study lounge space, plus a large number of computer stations and wireless internet access, thousands of students make use of the SLC each week. Students also visit the SLC to ask questions of instructors during their scheduled office hours that are held inside the center, to meet informally with fellow students to review course material, and to use SLC computers to complete online homework assignments and carry out other computer-based academic activities. However, beyond the operation of the SLC as a dynamic physical space that is conducive to these sorts of activities, SLC staff members also coordinate an array of peer-based learning opportunities through its large-scale PLSG program.

The PLSG program peer-based is a popular and highly valued option that has been intentionally structured to engage students actively and to foster collaborative learning outside of the classroom. It is well documented that collaborative learning can significantly enhance learning. For example, the effects of small-group collaborative learning on undergraduates in science, mathematics, engineering, and technology was examined through a meta-analysis undertaken by Springer, Stanne, and Donovan (1999), concluding that small-group learning has a significant beneficial effect on students' achievement, attitudes, and persistence. More recently, similar effects were noted in a seminal review of research evidence undertaken by Pascarella and Terenzini (2005). They concluded that effective collaborative learning environments allow students to experience learning in an active, engaged manner. Collaborative learning has also consistently been shown to increase student motivation and retention, leading to increased success. (Zhao and Kuh 2004; Kuh, Kinzie, Schuh, et al. 2005).

NSES

The PLSG program, employing over 250 high-achieving undergraduates each term to serve as skilled peer leaders to facilitate nearly 300 groups meeting weekly, helps to meet several goals set out as part of the National Science Education Standards (NSES). Although originally framed with K–12 education in mind, the NSES clearly has much to contribute to the discourse about how to foster a high quality postsecondary science education. Consider how the NSES teaching standards call for *more emphasis* on "sharing responsibility for learning with students" and also for "supporting a classroom community with cooperation, shared responsibility, and respect."

Students who opt to participate in the PLSG program are expected to take responsibility for their own learning by showing up each week fully prepared to actively engage with other group members and with the course content. Study group membership is voluntary, but in order to remain a member in good standing, students are required to attend regularly and participate actively. Members cannot miss two or more consecutive sessions or they may be dropped from the group roster.

The PLSG study groups are associated with natural science and math courses ranging from introductory biology and general chemistry to genetics, organic chemistry, and introductory physics. The groups are structured to provide a dynamic opportunity for students to study and interact with their peers in a cooperative and respectful environment. The meetings are designed to engage all group members, and are not meant to be tutoring or review sessions led solely by the peer leader. Leaders use facilitation techniques, encouraging and supporting members to teach and learn from each other. During their weekly meetings, peer leaders guide members to study and tackle challenging course concepts with the same group of peers. Members receive guidance and support from the trained study group leader who has already successfully completed the course, while at the same time group members build meaningful relationships with peers who are learning the same challenging material. Students who participate in the PLSG program are encouraged to develop a conceptual understanding of material previously presented in a traditional lecture setting and to apply concepts learned to new problems and situations.

At the UM, without an option for participating in the PLSG program, many students would experience their introductory science courses almost entirely through the large lecture sections that anchor such courses. Typically, during these lectures, instructors introduce and explain scientific content to the assembled group, expecting each student to learn the material by listening to what is presented and through reviewing the same material outside of class. Of course, spoken lecture material is sometimes supplemented by an occasional demonstration, the display of images or graphics, the posing of questions to the class by the lecturer, or through questioning of the lecturer by individual students. Also, lectures will occasionally be broken up by an in-class activity designed to divide the very large group down into smaller subgroups. However, in spite of these attempts to enhance the lecture hall environment, most learners spend a majority of their time during lectures listening to or viewing material without much opportunity to confirm their understanding or mastery of what's been covered. This is why it is particularly critical that PLSG members have a unique opportunity each week to assess their own understanding and expand upon it as a member of a small, supportive community of peers.

The NSES also urges educators to include students in the process of assessing their understanding, providing them with an environment conducive to scientific discussion and debate and focusing on understanding, knowledge use, ideas and the processes of inquiry. At the undergraduate science level, as in K–12 environments, learners benefit from increased emphasis in all of these same areas, which are supported by the PLSG program described in this chapter. Additionally, the PLSG program is built on a set of premises that tie into several other NSES content and inquiry standards, including reducing the emphasis on getting an answer versus increasing emphasis on "using evidence and strategies for developing and revising an explanation." During the PLSG program's leader training, study group leaders learn how to avoid teaching, tutoring, or giving answers, and instead how to use facilitation techniques to help study group members teach and learn from each other. As such, the peer leaders are responsible for setting the group agenda and making sure members stay on task. They are trained to recognize the importance of building a strong sense of community where all members feel comfortable taking intellectual risks and making mistakes as they engage in the struggle involved in building their own scientific explanations based on discussion, debate, evidence, and practice. The NSES recognizes that in the hands of a skilled leader, "such group work leads students to recognize the expertise that different members of the group bring to each endeavor and the greater value of evidence and argument over personality and style." Peer leaders provide the necessary skilled leadership as they are trained to use collaborative learning techniques and activities that encourage members to teach and learn from each other.

The NSES specifically addresses the importance of nurturing collaboration among students with the proclamation that "working collaboratively with others not only enhances the understanding of science, it also fosters the practice of many of the skills, attitudes, and values that characterize science" (NRC 1996, p. 46). What better way to do this for college-level science students than to provide an option for them to participate in collaborative groups focused on a "community of learners" model and led by thoughtfully trained peers? Peer leaders can also provide a student's perspective on course-specific requirements and expectations.

Details of PLSG Program

With the UM's introductory science courses enrolling as many as 1,800 students per term, it is not uncommon for enrollments in a single lecture to reach between 400 and 500 students. Some courses, such as general and organic chemistry and introductory biology, also offer smaller weekly discussion sections of 20–25 students as a supplement to the weekly lectures. These discussion sections are taught by graduate student instructors (GSIs) and are intended to provide students with opportunities to ask more questions and improve their understanding of the course material. Even so, these and the other large courses that do not offer discussion sections can present challenging learning environments for students, providing them with little opportunity to actively engage with the course material.

Research has long shown that student learning improves when there are opportunities to learn in small groups facilitated by peer leaders (Wamser 2006; Hockings, DeAngelis, and Frey 2008; Gafney and Varma-Nelson 2008). Therefore, the PLSG program was created nearly 15 years ago at the UM to provide undergraduates enrolled in introductory science courses with an opportunity to participate in such small groups. Participating in a study group provides many benefits. It presents students with structured weekly study time, which helps prevent procrastination and encourages study on a more consistent basis. It also increases the probability of exposure to the material in different ways that are better matched to a variety of learning styles. In study groups students also have multiple opportunities to teach each other what they know, and possibly identify weaknesses in their understanding through listening to other members' perspectives on the material. By teaching and learning from each other, students are encouraged to actively engage with the material, leading to increased mastery of the course content, growth in critical thinking skills, and greater conceptual understanding (McKeachie et al. 1986).

During the fall and winter terms, the SLC offers approximately 280 study groups supporting the more than 20 different courses in chemistry, biology, physics, astronomy, and math. Each group is composed of up to 13 peers enrolled in the course. Students from different sections of the same course can be enrolled in the same study group. Study groups meet once per week for two hours, and are facilitated by a trained peer facilitator known as a study group leader. Prior to exams, leaders often hold an additional two-hour session to help students prepare. Considering that groups begin meeting about two weeks into each term, and with additional optional exam prep and review sessions offered by leaders, most groups meet between 13 and 15 times during a typical 15-week term.

Staffing

The PLSG program has three full-time staff members: a director, a student affairs program manager, and an administrative assistant. The program also employs approximately 260 undergraduates as study group leaders during the fall and winter terms, and about 40 during each of the smaller spring and summer terms. Most leaders facilitate one group per week, although a small number of leaders lead two or three groups. The director and student affairs program manager are responsible for the recruitment, training, and supervision of the study group leaders. The administrative assistant handles program inquiries, payroll, room scheduling, and other administrative tasks.

Leader Recruitment

Over the summer, a recruitment e-mail is sent to all students who in the previous year earned a B+ or better in one of the more than 20 courses supported by the PLSG program. This e-mail includes a link to the study group leader application, which requires applicants to report their courses, grades, and other relevant information. Only students who have taken the course at the UM can become a study group leader for that course. This ensures they are familiar with the course content, structure, exams, and so on. The program also makes contact with faculty members, asking for names of students they would recommend for the position and personal e-mails are sent to those students encouraging them to apply.

Based on the number of returning leaders from the previous term, a determination is made about how many new leaders are needed. Program staff members then review the applications, looking for students with the requisite grades who have applied for courses with openings, and these applicants are invited to participate in group interviews. Group interviews are conducted just prior to the start of the term. Interviews are conducted by the director and the student affairs program manager, who each interview a group of five applicants for 30 minutes, and then switch to interview the other group. Interviewers use a rubric to evaluate each applicant's response to four questions in the areas of program philosophy, organization skills, facilitation skills, and problem-solving skills. Each response is ranked on a scale of 1–4, with 1 = Poor, 2 = Developing, 3 = Proficient, and 4 = Outstanding. A separate score for interpersonal skills, which the program believes are critical, is doubled, so each candidate receives an overall interview score from 6 to 24. Then, program staff members select the top candidates for each course to fill the number of positions needed. Generally, successful candidates demonstrate a strong passion for the course material, a commitment to the PLSG program philosophy, and exceptional interpersonal and communication skills, as well as outstanding organizational and problem-solving skills. During the Fall 2011 term, 366 students applied for a study group leader position, 202 were interviewed, and 124 were hired.

Training

All new study group leaders are required to participate in a three-hour new leader training session before they lead their first study group session. The focus of the session is to provide new leaders with critical skills they will need to successfully facilitate the weekly two-hour study group sessions. During training, leaders work closely with other leaders of the same course to develop and plan activities, such as icebreakers, that create and maintain a sense of community among the group members. New leaders also work in small groups with experienced leaders to define and discuss facilitation techniques meant to promote discussion among the study group members. Without simply "giving the answer" to member questions, new leaders also role-play to get practice at drawing out the knowledge of their study group members, discerning their level of comprehension, and guiding members who are confused. Study group leaders also work together to plan an agenda and discuss what activities to use during their first sessions.

Study group leaders also receive training in setting up their group space to maximize group interaction and learning opportunities. For example, they learn the importance of moving rows of desks into circles or a U-shape in order to give students better opportunities to discuss and collaborate. Study group leaders also learn to position themselves away from the whiteboard or chalkboard and to instead sit among the group members to emphasize their role as a facilitator rather than a teacher or tutor.

Additional training is provided to all leaders during collaborative course meetings held three times each term. Facilitated by a course leader who is a proven, skilled leader with experience facilitating groups for that course, these meetings provide a forum for leaders to collaborate on the development of weekly agendas, supplemental resources, discussion questions and activities, as well as talk about challenging situations in their study group. A PLSG staff member attends the meetings and provides short, targeted training sessions on topics such as collaborative learning techniques, learning games and activities, and dealing with difficult group situations.

Scale and Scope of the PLSG

Although originally piloted with a smaller selection of courses supported with study groups, the PLSG program has evolved to support more than 20 different courses, each enrolling from 100 to 1,800 students. The current list of courses for which the PLSG offers peer-led study groups includes two semesters of introductory biology, animal physiology, neurobiology, genetics, general chemistry, two semesters of organic chemistry, biochemistry, physical chemistry, two semesters of non-calculus based physics, two semesters of calculus based physics, intro physics for the life sciences, astronomy, and an honors calculus course. The courses with the largest enrollments, such as general and organic chemistry, typically have multiple sections taught by different instructors, although students in these courses do take common exams.

To support so many introductory courses, and so many with very large enrollment ones, the PLSG program is also very large, each semester typically enrolling approximately 3,200 study group members across the approximately 280 study groups associated with these courses.

Participants

At the UM, about 72% of students who participate in SLC study groups belong to the College of LSA, with approximately 22% affiliated with the College of Engineering and 4% with the School of Kinesiology. The remaining students are split among five of the other 16 Colleges and Schools at UM. The Fall 2011 breakdown was typical of most terms, with 42.3% of study group members who were second-year students, 34.9% first-year students, 18.7% juniors and 4.1% seniors.

Fall 2011 data are also typical with regard to the gender breakdown of study group members, with about 60% percent women and 40% men. Overall course enrollments for the same term showed a 51% male and 49% female profile. In terms of racial and ethnic diversity, study group members reported their racial or ethnic background as 74% Caucasian, 22% Asian, 5% African-American, 4% Hispanic, 1.5% Native American, and 1.5% Native Hawaiian/Pacific Islander, levels which generally mirrored overall course enrollments.

About 21% of study group members participate in multiple study groups (for different courses, as members are limited to only one group per course) and about half of all study group members have participated in a study group during a previous term.

Registration

Beginning the second week of the term, students enrolled in one of the courses supported by the PLSG program may register online for a group. Study groups are offered weekday evenings, except for Fridays, from 4 to 6 p.m., 6 to 8 p.m., and 8 to 10 p.m., and in two-hour blocks on Sundays from 12 p.m. to 10 p.m. Students may register for a group at any point during the term. If all of the study groups for a particular course are full, or the student is not available at the day and time of study groups with openings, a student may join the waitlist for one or more groups. When an existing member of a group drops the study group, the next student on the waitlist is automatically added to the group.

Overall, over 40% of students enrolled across all of the courses supported by SLC study groups choose to join and participate in at least one study group. Among the individual courses, participation rates for those who opt to join a study group range from 11% of enrolled students in Astronomy 101 to 71% of students enrolled in Organic Chemistry.

Study Group Sessions

During a typical fall term, approximately 280 study groups meet each week. Study groups are assigned locations for the duration of the term, typically classrooms or small seminar rooms distributed in several building on the UM's central campus. Some study groups meet within the SLC main branch in small alcoves, or in small team rooms within the SLC's satellite location.

Study group leaders have autonomy over the content and pace of their study group sessions, but collaborate with other leaders and program staff and solicit input from their study group members. Working with faculty, the PLSG program arranges for access privileges so leaders may log in to the Ctools site for their course, which usually includes a copy of the course syllabus, lecture slides, study guides, and other course materials. (Ctools is the UM's Sakai-based course management system.) The syllabus can give leaders an idea of what students have covered in class that week, although some large courses may have four or five different lecturers and these different instructors may cover the material at different rates or arrange topics in a different order. Therefore, leaders are encouraged to e-mail their members each week to learn more about what material was actually covered in the various lectures. Based on this knowledge, leaders formulate an agenda and identify and develop supplemental resources to bring to the study group. Leaders are required to send their agenda to PLSG staff each week for review, and to email their members material that previews the next study group meeting.

Study group leaders usually use the first 10 minutes of the group time to focus on community building. Some leaders will have an informal chat, while others will facilitate icebreakers early in the term to help their members learn each other's names and become more comfortable with each other. The better members know each other, the more likely they will be to take intellectual risks such as going up to the board to work out a problem they do not fully understand. Study group leaders and members may also occasionally bring snacks or meals to the group in an effort to promote social interactions and community building. Leaders are encouraged to spend time each week focusing on community building rather than abandoning these efforts after only a few weeks.

Leaders also have access to a PLSG program Ctools site, which includes worksheets previously developed by study group leaders, practice exams, collaborative learning activities, icebreakers, and other supplemental resources. Leaders are encouraged to use and revise these existing materials or to collaborate with other leaders to create new materials and to then post them back to the Ctools site. These supplemental materials are often the basis on which the discussion in the study group is centered. Leaders often break up the study group into smaller groups of two to four and vary the makeup of the groups so that each group member has opportunities to work with every other member of the group. This small-group work encourages members to teach and learn from each other, to learn different perspectives on the material, and to have more opportunities to actively engage with the course material.

Examples of the kinds of collaborative learning techniques used by Leaders include activities like *Buzz Groups, Think-Pair-Share, Jigsaw*, and *Pass-a-Problem*. A *Buzz Group* activity is simply a technique where the larger group of members is subdivided into several smaller groups and given the opportunity to consider a topic, question, or problem for a specified time period, after which the smaller groups report back to the full group. *Think-Pair-Share* is an activity that asks students to reflect or write individually on a topic or question for a few minutes, and then to pair up with a nearby student to discuss their ideas. After the pairs have discussed, the facilitator solicits comments from each pair as a way to promote discussion among the entire group. *Jigsaw* is an activity that requires students to become experts in assigned content areas, such as a concept or chapter, and then to teach other members of the group. Within the context of study groups, students typically work in pairs or small groups and then teach their content to other pairs or groups. *Pass-a-Problem* begins when the Leader distributes several envelopes that have problems attached to the outside. Each pair or small group must work through the problem and put the solution inside the envelope. The last group to receive the envelope must review and evaluate all of the responses, to choose the one that is most correct, and to explain to the large group why that response was better than the others.

Other activities frequently used by study group leaders include games like *Jeopardy*, which uses the course material for the answers and questions, and the *Flyswatter Game*. *Flyswatter* involves posting different course concepts or ideas on pieces of paper around the room. After dividing the group into teams, the study group leader reads a question and representatives from each team must quickly search the room and slap the correct response with their flyswatter.

In addition to these kinds of structured activities, leaders also ask members to bring their own questions to study group and to pose them to their fellow group members. This encourages members to take ownership of the study group and to rely less on the study group leader and more on the other members. Members are also encouraged to meet informally outside of the once-per-week study group session. These informal sessions usually occur without the leader and also help encourage members to teach and learn from each other, rather than relying upon the study group leader.

UM PLSG Program Compared to Similar Programs

Many aspects of the PLSG program share a basic framework with some very successful peer-based collaborative learning models in use at other institutions, including Supplemental Instruction (SI), Peer-Led Team Learning (PLTL), and Northwestern University's Gateway Science Program, to name a few. Clearly, many institutions have had to adapt the structure of their peer-based study group programs to match their own campus culture, resources, and requirements. So, while similar in many respects, each of these collaborative learning models differs from one another in many significant ways. Given these differences, we believe that there are several aspects of the PLSG program that make it easier to replicate at many colleges and universities where there is an interest in implementing a peer-based study group program. Below are some details about several of the components of the PLSG Program that may make this model easier for some institutions to implement.

- All peer-based study group programs provide an opportunity for students to engage with course material in a small-group setting held outside of the regular larger lecture or discussion structure. However, at some institutions, participation in a small peer-led group is mandatory, while at other institutions, such as at the UM, participation is optional. Since study groups at the UM are associated with such a large number of different courses offered by seven different academic departments, it is not feasible nor realistic to require mandatory study group participation for all students across all supported courses.
- Most programs require peer study group leaders to participate in a training program that addresses areas such as teaching and learning strategies and group facilitation techniques. However, at some institutions, peer leaders must enroll in a credit-bearing course or

seminar in order to participate in leader training sessions. Often, the grade for this course depends on the student's performance as a study group leader. At the UM peer leaders do not enroll in an additional course. Instead, their service as a peer leader is structured as a part-time position, and they are paid on an hourly basis for the time spent on activities such as attending training sessions, leading a weekly group, creating instructional materials, communicating with group members, and otherwise preparing for leading the weekly meetings. A student's performance as a study group leader is monitored by PLSG staff members more as part of job performance determination, rather than as a graded academic endeavor.

- Many peer-based study group programs require leaders to attend all regularly scheduled class meetings of the course for which they are leading a group. The PLSG program does not require its leaders to sit in on the course for which they are leading a group, although they are required to have taken the course at the UM in a previous term.
- Some peer-based collaborative learning models are based on the premise that for a program to be effective there must be a high degree of integration between the department offering study groups and the department(s) offering the courses supported with study groups. This is not the case for the PLSG model, where the SLC, as a centralized learning center, offers study groups for courses that range across seven different academic departments.
- Many peer-based study group program models rely upon the active and regular collaboration between faculty instructors and the undergraduate peer leaders, often requiring weekly meetings and the co-creation of instructional materials. The success of the PLSG program model does not depend upon such activities. As is true at many other institutions, UM faculty must devote much of their time to preparing for and delivering instruction, engaging in research activities, mentoring students, and attending other associated disciplinary and departmental commitments. These types of activities leave them little time for other activities. If the operation of an effective peer-based study group program at the UM depended upon an intensely active faculty involvement, there would be no such program. Certainly, this is also true at many other institutions.
- Many study group programs require all groups for the same course to work with the same centrally developed materials each week. The PLSG program does not prescribe what materials to use, nor does it require that all groups cover the same material each week. Instead, leaders and members together determine what topics to cover each week. Then, leaders are encouraged to choose from a library of instructional materials developed in previous years by peer leaders and/or to create their own materials.

Evaluation

The PLSG program evaluates members' experiences in their study groups and the effectiveness of their study group leaders through start-of-term and end-of-term surveys, direct observations, and debriefing sessions. Start-of-term surveys are distributed at the end of the third or fourth study group session. Members anonymously complete the surveys online. Program staff members review all of the surveys, and then give feedback to the leaders so they may make adjustments to their group, as necessary. If any serious issues or problems are discovered through reviewing the survey data, the program staff will meet with the leader to explore viable solutions. Typically, member feedback on start-of-term surveys is very positive, with effective leaders receiving praise for their understanding of the course material, focus on conceptual understanding, and distribution of supplemental materials such as worksheets and practice problems. Some members who are dissatisfied with their group or leader typically cite wanting more resources and problems, changing the pace or format of the group, varying the activities in the group, and more time spent preparing for exams.

Fall 2011 results of the start-of-term surveys were typical of the results we've collected over the past six or more years. In terms of evaluating their leader's performance, 97% of members said their leader provided them with useful worksheets, practice exams, and other resources, and 95% of members indicated their leader came to the study group prepared and sat with the group members (rather than standing at the board lecturing). Ninety-three percent of members said their leader-redirected questions to the entire group when a member posed one directly to the leader. Sixty-three percent of members reported that their leader encouraged them to meet for additional sessions on their own outside the regularly scheduled time, and 80% said their leader encouraged them to work with different people in the group. A list of training areas to emphasize and reinforce is built, in part, based on what is learned from the study group members' responses to this start-of-term survey. During the collaborative course meetings, held each month for leaders as a forum for ongoing training, a training agenda is built combining topics identified through analyzing survey responses with additional areas designated by staff as critical to the success of the program.

During the term, PLSG program staff members also conduct observations of many study groups and their leaders, particularly those leaders who are facilitating study groups for the first time. Observations typically last about 30 minutes. At the start of the observation, a program staff member checks to see if the room is arranged in a circle or U-shape rather than rows, so as to provide an environment more conducive to discussion. They also check if the leader makes efforts to create and maintain a sense of community in the group by leading icebreakers, name games, or through chatting informally.

Observations also provide an opportunity to see where the leader is positioned in relation to the group. Leaders are trained to sit with their group as a way to emphasize their role as a peer facilitator, rather than standing at the board, which would signal they are in more of a teaching or tutoring role. Effective leaders also divide their study group members into pairs or small groups of three or four, and then spend time engaged with each small group while the others discuss and work independently. Problems commonly identified at observations include issues with dominant members who monopolize the discussion, very shy or disengaged members, lack of structure, an insufficient variety of activities, or a lack of a sense of community that has resulted in a limited amount of discussion and member engagement.

Debriefing meetings of about 30 minutes are held shortly after observations so that new leaders have the opportunity to reflect upon their performance and to strategize improvements with program staff. The debriefing is structured as a dialogue, with leaders given the opportunity to identify their own strengths and growth areas. The observer then provides additional feedback, typically using the "sandwich" approach of beginning with a positive remark, offering constructive feedback in the middle and ending with another positive.

One of the challenges for the PLSG program is tied to the inability of staff to observe all of the leaders. Currently only one staff member is dedicated to conducting observations, which is insufficient to cover the entire staff of about 250 leaders. For this reason, observations focus on leaders who are new and leading groups for the first time, leaders who are facilitating multiple groups, and leaders whose feedback from start-of-term or end-of-term surveys suggest the need for improvement.

Although much is learned through surveying group members early in the term, most of the evaluation data on the PLSG program comes from more comprehensive online surveys conducted at the end of each fall and winter semester. Leaders are asked to set aside some time during one of their final two study group sessions for members to complete the survey. This end-of-term survey asks members to report on why they joined a study group, whether and how they benefited, and what influence, if any, study group participation has had on a number of factors. As a result of having their own work critiqued and through the act of explaining concepts to others, study group members overwhelmingly report experiencing many benefits of study group participation.

The response rate for fall 2011 was 41.9% (1,436 responses out of 3,422 members). An analysis of these responses shows that over 92% of study group members would recommend their study group leader to a friend. Most study group members reported that their participation in a study group resulted in a number of positive effects with the highest percentage of members reporting an increase in their: understanding of difficult concepts (92.1%); confidence about their mastery of the course content (86%); belief that study group participation positively affected their exam grades (85%), as well as their overall course grade (80%).

Interestingly, members also indicated in significant percentages that their participation in a study group positively affected their decisions about taking more science classes and majoring in science. Forty-four percent of members indicated that based on their study group experience, they would be more likely to take additional science classes. Forty-nine percent of members said their participation in a study group made them more likely to maintain their major in a science-related area, while another 16% reported the study group made it more likely they would switch their major to a science related one. These findings support evidence in the literature showing that student persistence to degree in scientific disciplines is linked to the degree to which they participated in small peer-based academically oriented groups (Light 1992).

The PLSG program also conducts comprehensive end-of-term surveys of the study group leaders. In the fall 2011 survey, leaders cited a number of benefits from their work, including improved communication and facilitation skills, increased confidence and patience, improvement in their own study skills, and development of their leadership skills. One leader described the program's impact on him this way: "I think that taking part in the study group program has helped me to improve my facilitation skills and my interpersonal skills. I think that this semester (compared to last semester), I grew more comfortable interacting with my study group members (last semester I was too formal, in my opinion)."

Leaders also developed metacognitive understandings, which in turn improved their ability to think critically about the course material. For example, one leader stated the following: "I think I have discovered a lot more about how I learn and therefore how to help others understand concepts when they may learn differently." Another leader was able to discover the difference between grades and learning: "It made me even better realize what type of student I am and what my priorities are for learning. It has shown me that I do have my priorities in the correct place: the grade isn't the most important thing, learning the material is."

As with some members, there were leaders who also discovered their participation in the program helped them determine future graduate school and career plans. For example, one leader said, "I had always wished to go into education, but was convinced on being pre-med. After noticing my passion for helping others and the personal satisfaction I receive from helping students understand a concept, I am now a biology/education major."

What's Next?

Studies of programs similar to the PLSG have been conducted that have shown a positive correlation between participation in a collaborative learning program with increased learning and higher course grades (Varma-Nelson 2006; Springer, Stanne, and Donovan 1999). Although our students self-report that they have benefited in a variety of ways from their participation in the PLSG, we have not yet been able to carry out a rigorous controlled study on the effects of participation. A plan to do so is now underway, with a strategy to use as a control group those students who wished to register for membership in a study group but who, for reasons of scheduling or timing, were not able to get into a group before all available groups were filled. We also hope to determine whether any specific groups of students benefit more or less from their participation in study groups, such as better-prepared versus lesser-prepared students; or males versus females; or first-generation students; or students from groups typically underrepresented in the sciences; or students who attend more of the term's weekly sessions versus those who attend fewer, and so on.

The PLSG program at the UM serves a large number of students across a wide variety of science courses who resoundingly report that they have benefited from their participation. Peer leaders also report many positive outcomes related to their roles in the PLSG program. For these reasons we believe that the implementation of similar programs at a greater number of institutions will benefit science learners across the country. Adaptations may be made to fit institutions of varying sizes and with student populations that exhibit characteristics different from those attending the UM. It is clear that there is no one way to operate a highly effective peer-led study group program; we encourage adopters to make as many adjustments to the PLSG model, or to any of the other similar models, as are necessary to match the needs of their campus. As long as students are given the opportunity to meet regularly with the same small group of peers; are actively engaged in teaching and learning from each other; and are led by a peer who is familiar with the course content and who has been trained in community-building and group facilitation techniques, they will benefit.

References

Astin, A. W. 1993. *What matters in college: Four critical years revisited*. San Francisco: Jossey-Bass.

Barkley, E. F., K. P. Cross, and C. H. Major. 2005. *Collaborative learning techniques: A handbook for college faculty*. San Francisco: Jossey-Bass.

Boyer Commission on Educating Undergraduates in the Research University, S. S. Kenny (chair). 1998. *Reinventing undergraduate education: A blueprint for America's research universities*. Stony Brook, NY: State University of New York-Stony Brook.

Gafney, L., and P. Varma-Nelson. 2008. *Peer-led team learning: Evaluation, dissemination, and institutionalization of a college level initiative*. New York: Springer Science+ Business Media.

Hockings, S. C., K. J. DeAngelis, and R. F. Frey. 2004. Peer-led team learning in general chemistry: Implementation and evaluation. *Journal of Chemical Education* 85 (7): 990–996.

Kuh, G., J. Schuh, E. Whitt, and Associates. 1991. *Involving colleges: Successful approaches to fostering student learning and development outside the classroom*. San Francisco: Jossey-Bass.

Light, R. J. 2001. *Making the most of college: Students speak their minds*. Cambridge, MA: Harvard University Press.

McKeachie, W. J., P. R. Pintrich, Y. Lin, and D. A. Smith. 1986. *Teaching and learning in the college classroom: A review of literature*. Ann Arbor, MI: University of Michigan.

Pascarella, E. T., and P. T. Terenzini. 1991. *How college affects students: Findings and insights from twenty years of research*. San Francisco: Jossey-Bass.

Pascarella, E. T., and P. T. Terenzini. 2005. *How college affects students, volume 2: A third decade of research*. San Francisco: Jossey-Bass.

Springer, L., M. E. Stanne, and S. Donovan. 1999. Effects of small group learning on undergraduates in science, mathematics, engineering and technology: A meta-analysis. *Review of Educational Research* 69: 21–51.

Wamser, C. C. 2006. Peer-led team learning (PLTL) in organic chemistry: Student performance, success, and persistence in the course. *Journal of Chemical Education* 83 (10): 1562–1566.

Zhao, C., and G. D. Kuh. 2004. Adding value: Learning communities and student engagement. *Research in Higher Education* 45 (2): 115–138.

Take Your Students Outside: Success With Science Outdoors

Beth Ann Krueger
Central Arizona College–Aravaipa Campus

Introduction

"This is why I love science," exclaimed Magdelena (not her real name), splashing back streamside with a pan full of macroinvertebrates she had just collected. "This is so FUN!" She proceeded to record the number and types of each macroinvertebrate with the help of two classmates. Magdalena was a student in the author's general biology (BIO 181) course at Central Arizona College's Aravaipa Campus. She and her classmates conducted field research at The Nature Conservancy's San Pedro River Preserve and the Aravaipa Preserve, both near Winkelman and Mammoth, Arizona. The focus of the course project was water quality assessments of the San Pedro River and Aravaipa Creek. When the author first started this project over five years ago, these students were one of the very few Arizona community college freshmen participating in actual field research as a part of their class. Now, happily, this is no longer the case.

Setting

Central Arizona College (CAC) is the community college serving Pinal County, Arizona. Pinal County has a diverse population both in terms of demographics (see Tables 5.1 and 5.2, p. 62) and geography. Four Native American reservations are found in Pinal County: Ak-Chin Indian Community; Gila River Indian Community; Florence Community (Tohono O'Odham) and San Carlos Apache Indian Reservation. Several mountain ranges run generally north/south through Pinal County: the Galiuros, Santa Catalinas, Tortolitas, Minerals, Superstitions, and Sacaton Mountains. One of the highest peaks in southern Arizona, Mount Lemmon (9157 ft.; 2791 m), is located in the Santa Catalina Mountains in northeastern Pima County, near the border of Pinal County. Three rivers are found (but often don't flow) in Pinal County: the Gila, Santa Cruz, and San Pedro. Five state parks, including Picacho Peak and Lost Dutchman and various wilderness and protected areas, such as the Aravaipa Wilderness, The Nature Conservancy preserves, and the Coronado National Forest, are also found in Pinal County.

Figure 5.1. Students learn from TNC Biologist Celeste Andresen (in middle, with hat) how to operate monitoring equipment. Delilah Carbajal is on the right (white t-shirt).

Table 5.1. Age Demographics of Pinal County, Arizona (CAC Fact Book, Spring 2012)

Under 20 years	29.2%
20–39 years	27.7%
40–59 years	23.4%
60–79 years	17.2%
80 years and more	2.5%

Table 5.2. Race Demographics of Pinal County, Arizona (CAC Fact Book, Spring 2012)

American Indian or Alaskan Native	4.5%
White–Hispanic	23.3%
White–Non-Hispanic	58.5%
All other races	8.6%
Two or more races	5.1%

Accredited by the Higher Learning Commission of the North Central Association of Colleges and Schools, CAC is composed of three campuses and seven centers located throughout Pinal County. For Spring 2012, total headcount was 6,707, with 2,041 full-time and 4,666 part-time students. Females accounted for 64.5% of the student population and the average age of a CAC student for this semester was 29. CAC offers over three dozen associate degrees and certificates ranging from business to nursing to fine arts to medical transcription. Other nonacademic programs include College for Kids, a summer science program at the Aravaipa Campus; Lifelong Learning classes, for example Introductory Painting Skills and English as a Second Language (ESL) classes. Finally, courses are offered in traditional classroom "face-to-face" format, online, via interactive television (ITV) and in any combination of these three, called "hybrid" format.

Central Arizona College's Aravaipa Campus, where the author teaches, is situated in far eastern Pinal County. This area of the county once had active copper mining. However, only one mine (ASARCO Ray Mine) and associated smelter in Hayden, Arizona, are active at present. The BHP San Manuel Copper Mine, (San Manuel, Arizona) and smelter permanently closed in 2003. Therefore, the Aravaipa Campus provides critical job training, associate degrees, and university transfer programs for area residents. In a transfer partnership with Northern Arizona University (NAU), students can even obtain a bachelor's or master's degree in some fields (for example, administration). NAU representatives have an office on campus and visit at least several times a month.

For Spring 2012, the Aravaipa Campus headcount by physical location was 193 students. Average course size was about 10 students. Fourteen students graduated from this campus in Spring 2012. Four full-time faculty (math, science, social sciences, and communications/English) are present. Four ITV rooms, a library and learning center, and student services offices also serve the campus. In addition, one building is devoted to technical and trades training (auto body and carpentry, for example).

Figure 5.2. Student Records a GPS Reading

The Nature Conservancy (TNC) San Pedro River Preserve, TNC Aravaipa Preserve, and TNC managed 7B's Ranch (owned by Resolution Copper Mining) are located within 10–15 miles of the Aravaipa Campus. The author has formed a partnership with TNC staff so that

Aravaipa Campus students in her science classes can do real field research including water-quality studies, bird surveys, and more alongside TNC scientists.

Several CAC science professors (engineering/physics; geology/astronomy; and chemistry) are striving to incorporate National Science Education Standards (NSES) goals into the curricula and to get their students outside on field trips and engaged in critical thinking-oriented (as opposed to typical "cookbook") labs. This chapter will focus on the biology and environmental science projects that the author has been developing in partnership with the TNC. The target learners are biology, environmental science, and chemistry students as well as nonscience majors at the Aravaipa Campus, which, as mentioned, is located in the beautiful Sonoran Desert near the San Pedro River and Aravaipa Creek riparian areas. There is no town in this area, only the campus and some scattered houses, farms, preserves, and ranches.

The importance of connecting biology and science students to real-world issues and the natural environment has been discussed in numerous journal articles, books, and education conferences and meetings (Neilson 2009: Nisbet, Zelenski, and Murphy 2009; Simms 1996; Zita 2008). Most well-known of the literature, perhaps, is Richard Louv's book *Last Child in the Woods* (2005) in which Mr. Louv called attention to "nature-deficit disorder" in children. Getting students involved outside the classroom is also part of the National Science Education Standards:

> IDENTIFY AND USE RESOURCES OUTSIDE THE SCHOOL. The classroom is a limited environment. The school science program must extend beyond the walls of the school to the resources of the community… (NRC 1996, p. 45)

Saddened by the disconnect many of her students have with nature, the author is continually trying to find ways to incorporate outdoor and "real-life" experiences into classes; therefore, she started the project described here.

Figure 5.3. TNC scientists (foreground, wearing hats) look on as students take readings from a pisometer near the San Pedro River.

Major Features of the Instructional Program

While the author had included field trips; mini-research projects; and outdoor tours of parks, botanical gardens, and natural areas into her classes, it was the Arizona Rivers Project which introduced her to the process of incorporating actual semesterlong field studies. Soon after moving to Arizona, she attended a 3-day workshop in Phoenix. More than 20 educators (K–12 teachers, 4-H directors, education directors, and one community college teacher) participated in various field exercises, using equipment and getting familiar with established protocols that they would then take back to use in their schools. The focus, based on the mission of the Arizona Rivers group, was water quality. Participants learned how to test and monitor water quality factors (pH, conductivity, presence/absence of various substances, invertebrate surveys, and hydrological measurements). Participants then took these protocols and procedures back to their classrooms, 4-H project workshops, and education centers, and incorporated them into their science programs.

One major challenge for teachers is that much of the scientific equipment available to schools has one or more of these issues: a large learning curve (discouraging students and wasting valuable lab/field time); expensive to buy and maintain; and cannot take the rigors of field use (it may require a computer, for instance). One important concept emphasized by the Arizona Rivers Project training staff was that quality field work could be done by students with inexpensive, user-friendly, field-friendly equipment. Examples included handheld pH and conductivity monitors that could float and inexpensive water test kits from biology supply stores.

Another advantage to the Arizona Rivers approach was that they used standard protocols, such as those from GLOBE (Global Learning and Observation to Benefit the Environment; *http://globe.gov*). In addition, if students and teachers chose to upload their data to the GLOBE website, they could share and compare their observations with those from other student groups around the world. They are doing real research and comparing and sharing data with colleagues!

Collaboration With Research Organizations

Once the author completed the Arizona Rivers training, she needed to find local people and organizations with which to partner. Again, Aravaipa Campus is located in "the middle of nowhere"; the nearest towns are about 10 miles away, and these are small communities (Mammoth, population about 1,400; Dudleyville, population about 950; and Winkelman, population about 350). The average number of students at this campus is about 150–200. While this may be a challenge for enrollment, it was an advantage for the author, who wanted to do fieldwork. TNC San Pedro Preserve, the TNC Aravaipa Creek Preserve, and the TNC-managed 7B's Ranch property are all located a short drive away. The author contacted TNC staff, outlined her ideas for student projects at a meeting, and consequently the staff was eager to become a partner in the project. After several more logistical and organizational talks, e-mails, and meetings, appropriate student activities were agreed on that would meet the science and course education goals, and would also help the TNC gather necessary data.

At the beginning of the class, students were introduced to the concept of a desert riparian ecosystem, the importance of water quality and use, and other related subjects by reading "Desert Waters: From Ancient Aquifers to Modern Demands" by Nancy K. Laney (1998) and doing online research specifically about the San Pedro River and Aravaipa Creek. They were also instructed by a library staff member in proper research techniques and in how to determine the credibility of a resource. Next, they were broken randomly into research groups. Three 3-hour lab periods per semester were devoted to actual research at the preserves and three 3-hour lab periods were devoted to working together in their group to analyze the data, draw conclusions, and formulate recommendations. The final evaluation was a report for the first year this project was done. For the second year, a formal research paper, complete with references, data tables, and such, was required.

Figure 5.4. Students Take Readings From Another Pisometer Near the San Pedro River.

Evidence for Successes

The first year, only a very small number of students were enrolled in the general biology class, but they were enthusiastic about doing fieldwork as part of the class requirement. Water-quality monitoring, which included use of the handheld monitors and water kits mentioned earlier, was the primary focus for this group. In addition, students learned to use some complicated TNC hydrological equipment as well as a pisometer (tool used to measure water depth) stationed along the San Pedro River. At the end of the semester, they compiled their data in a report and submitted it to the TNC staff. An excellent result of this was that one of these students, because she had participated in this class research project, was hired by the Bureau of the Interior on a part-time basis to assist researchers with small mammal surveys along the San Pedro River. The following semester, she and other students who continued with the second semester general biology class, participated in plant and bird surveys and other research studies at the TNC in addition to doing more water-quality analysis.

Figure 5.5. Students collect macroinvertebrates on the NC San Pedro Preserve. Monique Boger is the student on the right (black shirt).

The second year, a larger class (about 12 students) enrolled in general biology. In addition, students had the opportunity to visit both TNC Aravaipa Creek Preserve once and TNC San Pedro Preserve several times during this year. The focus for this group was again water-quality analysis. Class and field assignments were revised and updated from the previous year in conjunction with TNC staff. For example, as mentioned, a formal research report was required at the conclusion of the project. This report was also shared with the TNC scientists, who again used the students' data and observations. About half of that class went on to complete a rigorous one-year anatomy/physiology sequence and are now applying for or attending nursing school. Below are student comments about the project from students who have continued in the sciences/health career fields of study and also commentary from one of the TNC scientists:

Monique Boger (Figure 5.5): I really enjoyed taking the biology class and doing the River Project. It really helped me understand better how to work well with a group and how to put together an actual group research project. I enjoyed being in the field and being able to identify all of the different macroinvertebrates, fish, and plants that we would come across. Being able to be out in the field gave me a greater appreciation of the outdoors; knowing a little bit more about the trees and animals that live out there and the importance of water quality.

Delilah Carbajal (Figure 5.1, p. 62): Studying science was at times a challenge, but it eventually helped me to become a much better student. Working outside the classroom in the field was a great experience. Comparing the different macro-invertebrate specimens as part of a water-quality study taught me that species diversity is an important clue to environmental change. I enjoyed the visual aspects of doing fieldwork: Observing endangered fish and actually being in a desert riparian system proved to be most helpful in the field verses the classroom. Actually smelling the foliage around the river and seeing the imprints of the different bird species made this experience one I won't forget. Working alongside TNC scientists who can actually explain what we were learning about in understandable terms was another bonus. This was a great experience and I would encourage any students or schools and colleges to consider the ongoing importance of these programs, as it helps in many different areas of learning and allows student to participate in real research.

Figure 5.6. Lowland leopard frog at a wetland (near the San Pedro River) monitored by the Nature Conservancy.

This is a "species of concern" because its habitat is disappearing and invasive species (bullfrogs) prey on it. Student research on water quality in this wetland is used by TNC biologists.

Michelle Jackson: Working on the river project under the instruction of Dr. Krueger was one of the most memorable moments of my college career. Analyzing the water, identifying the numerous macroinvertebrates and plants present, and being able to learn and experience it all through the eyes of scientists and my teacher made the experience that much more enjoyable, educational, and memorable. Every day that we were at the river was an adventure as well as a journey; having the TNC scientists and Dr. Krueger there to explain, educate, and encourage was what made the project exciting. I learned so much about the environment where I live; all this was literally in my backyard. There were challenges, but with patience and the continual encouragement from my teacher, I overcame those obstacles; looking back, it was so informative, fun, and definitely an experience that will stay with me forever.

In the past two years, this project was put on hold for two very valid reasons: (1) the author was selected to teach in China for six months in 2011; and (2) enrollments have not allowed the college to offer the general biology or environmental science classes at the Aravaipa campus. The good news is that for fall 2012, a general chemistry course was scheduled and a student research project on the TNC-managed 7B's Ranch property was planned as part of the course.

Finally, Celeste Andresen (Figure 5.1 and 5.8), TNC scientist and manager of the 7B Ranch property, stresses the importance of this project to the community, the river, and the TNC (personal communication 2012):

The San Pedro River flows out of Mexico, north into Arizona. As one of the last undammed rivers in Arizona, it is a vital migratory flyway and wildlife corridor to many species of fauna. The river provides intact cottonwood-willow galleries, and is the location of one of the largest mesquite bosques in the desert southwest.

The Nature Conservancy has had a presence on the River and its tributaries for several decades. This discussion focuses on the Lower San Pedro River, which begins near Cascabel, Arizona, and includes the reach from "The Narrows" north to the confluence with the Gila River outside of Winkelman, Arizona. TNC owns and manages properties adjacent to the San Pedro River and association tributaries. Hydrological data is collected throughout the San Pedro Watershed on a monthly basis. Wet/dry Mapping is performed annually to determine which reaches have perennial flow. The data that is collected is critical to supporting the complete San Pedro Watershed story.

Dr. Beth Krueger and her students have been instrumental in the data collection process on the Lower San Pedro River Preserve (TNC), West Aravaipa Preserve (TNC), and 7B Ranch. The 7B Ranch is a property owned by Resolution Copper Mining, and managed by TNC staff. Dr. Krueger includes field trips with her class to perform water chemistry analysis, stream flow measurements and depth-to-groundwater monitoring. The data are collected, and then entered into the TNC database where it is available to TNC staff for a variety of uses, such as tracking trends, or inclusion in reports or journals. The students' participation in the collection of data is a vital part of completing the hydrologic story of the San Pedro Watershed.

Figure 5.7. Student David Rowlands conducts water quality research at the wetland.

Looking Ahead

Further consultation with TNC scientists will allow expansion of class projects into new areas of study that best fit TNC needs as well as the student learning objectives of the particular courses. For example, as stated, the 7B's Ranch property manager is in need of some basic water chemistry/water monitoring studies, therefore the author is working with her on the details of this project for the fall 2012 Aravaipa Campus general chemistry class. Eventually, with enough research data, students could even participate in student research symposiums at the college and in the region upon the completion of their class.

Basic Considerations Before Incorporating Field Research

If one is planning to incorporate this type of project into a class, it is recommended that the following items be considered first:

- Proximity of organizations that are willing to host a class (Nature Conservancy, for example)
- Obtaining teacher training, if needed; training can be formal or informal
- Funding and administrative support
- Equipment and sites available
- Transportation of students
- Student release forms

This list is not comprehensive, but it does cover the bottom-line items that need to be addressed before incorporating fieldwork into the curriculum.

First, find organizations in the area that are involved with field research, even if the research is on a small scale. Organizations could be local chapters and/or sites of national/international organizations, such as Audubon, the Nature Conservancy, and so on. State parks, national parks/ monuments, and local nonprofit parks and museums (such as botanical parks) are also potential places to investigate, especially since many have and encourage education outreach programs. Sometimes universities are not especially amenable to including outside groups, such as a high school or community college classes, in their field research projects; however, one could also investigate this avenue to see how their local university responds. Finally, small-scale projects could be done right on school grounds. For example, an elementary school in my area, Mammoth Elementary STEM School, has planted a garden that includes native crops and also a pollination garden. Students study the importance of native animal species to plant pollination. Students also complete science fair projects, including one addressing how soil affects plant growth. At the 58th Annual Southern Arizona Regional Science and Engineering Fair, the school was selected as the top K–5 elementary school (Copper Area News Publishers 2012).

The next step is to ensure one's supervisor and the administration will support the project. Issues to consider include cost of travel to the field site, equipment cost (see below), time, and faculty training needed. Depending on the location and equipment available at the field site, the first two issues may not be problematic. The Aravaipa Campus has a van that can carry up to 12 students to the TNC field trip sites, which are less than 15 miles away. Therefore, gas mileage costs to the department are quite small (less than $100 per semester). Students are also allowed to meet the class at the site with permission from the instructor and TNC.

Figure 5.8. TNC Biologist Celeste Andresen (right foreground, white ball cap and shirt) assists students with water-quality monitoring and macroinvertebrate collection on TNC San Pedro Preserve. Michelle Jackson is the student on the left (foreground, in ball cap and polo shirt).

If opposition is encountered, one way to convince recalcitrant administrators is to have people in the community (parents and fieldwork partners, including students) support the project. Have them request that fieldwork be incorporated into the curriculum. In addition, collecting current examples of student fieldwork from other schools and colleges will provide positive evidence for the project. It is also important to make connections with other teachers (such as at the National Science Teachers Association national conference) who will be able to give excellent advice about gaining both administrative and monetary support for the project. Making connections with other teachers can also help one locate opportunities for teacher training.

Limited (or no) budget is a severe issue. Applying for small grants for teachers can be especially helpful with the purchase of equipment. The Arizona Rivers Program had applications for $300 grants; teachers who participated in the Arizona Rivers workshop could apply for these small grants. The Aravaipa Campus was able to obtain some equipment in this way. Some grants are actually awarded directly to the applying person and not to the school, which enables the teacher to take the equipment with them if they move to another school or college. Also, occasionally local businesses will support field projects with either cash or in-kind donations or the use of their land for a small garden or other science project.

The field equipment does not have to be fancy or cost a lot. I use readily available water treatment test kits (available from biological science teaching companies such as Wards and Carolina) and handheld pH meters, conductivity meters, thermometers, and so on. The handheld monitors cost about $100 each (significantly less than a traditional benchtop pH meter and a lot less maintenance). These handheld monitors have an additional advantage in that they float. Other equipment you can make yourself. For example, instructions for making a Secchi disk can be found online (*http://des.nh.gov/organization/divisions/water/wmb/vlap/documents/secchi.pdf*). Another example is to make quadrants for plant surveys (the Aravaipa Campus has ones made of lightweight PVC pipe) in-house. There are numerous websites available with instructions on how to construct simple low-cost field equipment.

For teachers who want more training in fieldwork, it is recommended that they find classes specifically for teachers, held by organizations such as the Arizona Rivers Project. Global Learning and Observation to Benefit the Environment (GLOBE; *http://globe.gov*) has a teacher workshop website (*http://classic.globe.gov/fsl/workshop/registration.pl?&lang=en*). Other options include volunteering with organizations that welcome citizen scientists (National Weather Service Weather Spotters Program; Bird Feeder Watch–Cornell Lab of Ornithology; local parks and museums, and so on). These organizations often have free or low-cost professional development for their volunteers. Finally, there are formal courses, such as those offered by Montana State University (MSU), designed for teachers to improve their understanding of a subject as well as how to apply that knowledge to class lessons and projects. For example, the author enrolled in an online graduate course in Grasslands Ecology at MSU. As part of the class, simple field techniques were taught that could be used by undergraduate first- and second-year students. The final project for this Grasslands Ecology course required devising a semester field research lesson for students, and then doing the experiments for the project using some of the techniques learned in the class.

Summary

Incorporating field research into one's classes is critical in helping students to gain an appreciation of nature and the affect of human activities on the environment. What better way to meet science education goals than by having students outside doing science?

Figure 5.9. Spring 2013 Students With Dr. Krueger (far right in large hat) After a Morning of Water-Quality Studies at the Wetland

References

Andresen, C. 2012. Personal correspondence.

Copper Area News Publishers. 2012. Mammoth STEM students have great show at science fair. *Copper Basin News*. March.

Laney, N. K. 1998. *Desert waters: From ancient aquifers to modern demands.* Tucson, AZ: Arizona Sonoran Desert Museum Press.

Louv, R. 2005. *Last child in the woods: saving our children from nature-deficit disorder*. Chapel Hill, NC: Algonquin Books.

National Research Council (NRC). 1996. *National science education standards*. Washington, DC: National Academies Press.

Neilson, A. L. 2009. The power of nature and the nature of power. *Canadian Journal of Environmental Education* 14: 136–148.

Nisbet, E. K., J. M. Zelenski, and S. A. Murphy. 2009. The nature relatedness scale: Linking individuals' connection with nature to environmental concern and behavior. *Environment and Behavior* 41 (5): 715–740.

Simms, K. 1996. Geostudies: Structuring a multi-credit outdoor environmental course. *Green Teacher* 49: 19–22.

Zita, A. 2008. Technology works in the outdoors. *Pathways: The Ontario Journal of Outdoor Education* 20 (2): 8–10.

Resources for Teachers

Central Arizona College Facts Book. *www.centralaz.edu/Home/About_Central/ Institutional_Planning_and_Research/Fact_Book.htm*

Citizen Science Alliance. *www.citizensciencealliance.org*

Cornell Lab of Ornithology. *www.birds.cornell.edu/Page.aspx?pid=1478*

Cornell Lab of Ornithology Citizen Science. *www.birds.cornell.edu/citsci*

Bird Sleuth Curriculum. *www.birds.cornell.edu/birdsleuth*

GLOBE: Global Learning and Observations to Benefit the Environment. *http://globe.gov*

Hanna Instruments. *www.hannainst.com/usa*

Montana State University—National Teachers Enhancement Network. *http://btc.montana.edu/courses/aspx/ntenhome.aspx*

National Association of Biology Teachers (NABT). *www.nabt.org*

Student Perspectives on Introductory Biology Labs Designed to Develop Relevant Skills and Core Competencies

Ellen H. Yerger
Indiana University of Pennsylvania

Setting

The Pennsylvania State System of Higher Education (PASSHE) has 14 schools across the state, amongst which Indiana University of Pennsylvania (IUP) is the largest. IUP is a teaching/research/doctoral university that offers more than 140 undergraduate majors and over 70 graduate programs. Faculty research, publish, and consult in their fields while focusing their energy in the classroom.

Students come from a wide range of high school programs, reflecting their rural, suburban, or urban character. There is a wide range of preparation, motivation, and interest among the 15,000 students studying at IUP.

Overview

Several decades ago as president of the University of Minnesota, Meredith Wilson urged educators to conceive of the students as the greatest energy source available on campus (Wilson 1967). He noted that there are at least 10 times as many students as teachers, and if we conceive of students as supplicants gathering to be taught, we are left with only a small population of teachers as the active agents. If, instead, we regard students as learners, they become the essential engines of their education, and we unleash 10 times as much intellectual power.

Even more so today, we recognize that students are the active agents for successful learning. The National Science Education Standards (NSES) emphasize empowering students to share responsibility for learning. To make this happen, the standards envision teachers guiding students in active and extended scientific inquiries that focus on the *use* of scientific knowledge, ideas, and content (NRC 1996).

This approach is also highlighted in a landmark document, *Vision and Change in Undergraduate Biology Education: A Call to Action*, supported by the American Association for the Advancement of Science (AAAS) and the National Science Foundation (NSF). Brewer and Smith (2009) state that

> the practice of biology requires more than just understanding core concepts. To understand, generate, and communicate knowledge about the living world, students need to develop and apply relevant skills. Therefore, in addition to understanding concepts, undergraduates must have opportunities to develop core competencies to better prepare them to practice biology, as well as to address the complex biology-related issues that our society faces. (p. 11)

With these challenges in mind the biology laboratory curriculum at IUP has been created. The intention is to develop core competencies and relevant skills in our first-year introductory biology students that will grow with them as they go through upper-level courses and into their careers.

Major Features of the Introductory Biology Lab Program

The biology department at IUP has been teaching introductory labs for many years. The labs have been developed over time as new ideas are contributed by the many faculty teaching sections of the course. This year a group of several faculty gathered the most successful ideas, reflecting what is done well, and identified the recurring challenges. The labs were modified by faculty consensus to address these common issues:

- Reinforce difficult core concepts with hands-on learning.
- Teach relevant skills that students can further develop and apply.
- Focus on the most important core competencies students need, with less concern for broad coverage of content.
- Teach techniques for independent learning and research.
- Prepare students to succeed in later courses for which this course is a prerequisite.
- Be interesting and useful to all students, having a wide range of interests, motivation, and preparation.
- Be accessible to students underprepared from high school.
- Stay within a limited budget.
- Value limited faculty time.

Evidence for Success

Type of Information Collected

Whatever the appropriate criteria for judging teaching, the foremost question is how were the student learners affected? A biology lab survey was given to the students at the end of the semester asking them to evaluate the labs (Figure 6.1). The students ranked all the labs as 1st (most), 2nd, 3rd, down to 10th (least) on these survey questions.

Figure 6.1. Lab Survey Given to Introductory Biology Students at the End of the Semester

Name ______________________________

First complete the rankings for questions 1–5 in the tables below.
After fully completing the tables, answer the 2 questions that follow the second table.
Thank you.

1. Rank all 10 labs. Which labs helped you to learn the most biology facts and concepts?
2. Rank all 10 labs. Which labs helped you to learn the most lab skills?
3. Rank all 10 labs. Which labs were most interesting to you?

	Rank 1 for most and 10 for least		
Labs (in the order done during the semester)	**Your Ranking for Question 1**	**Your Ranking for Question 2**	**Your Ranking for Question 3**
Intro to the Scientific Method. Generating and testing hypotheses from data sets.			
Microscopy. Using a microscope, observing cells with a microscope.			
Water Relations. Osmosis in potatoes exposed to various sucrose concentrations.			
Spectrophotometry. Beet tissue homogenate absorbance spectrum and standard curve.			
Spectrophotometry. Using a standard curve to assay beet tissue exposed to acetone and methanol.			
Enzymes. Acid phosphatase assays of various plant tissues.			
Mitosis and meiosis. Microscope observations of cells dividing, and paper diagrams.			
Mendelian Genetics. Corn monohybrid and dihybrid crosses, Chi square tests, problems.			
DNA fingerprinting and forensics. Restriction enzyme cuts of DNA, and gel electrophoresis of DNA fragments.			
Plasmid DNA transformation of bacteria, agar plating, Glo gene expression.			

4. Rank all 10 labs. Which labs had the most difficult concepts to understand?
5. Rank all 10 labs. Which labs required the most time to complete, when including time spent outside of lab, either to understand it or to write it up?

	Rank 1 for most and 10 for least	
Labs (in the order done during the semester)	**Your Ranking for Question 4**	**Your Ranking for Question 5**
Intro to the Scientific Method. Generating and testing hypotheses from data sets.		
Microscopy. Using a microscope, observing cells with a microscope.		
Water Relations. Osmosis in potatoes exposed to various sucrose concentrations.		
Spectrophotometry. Beet tissue homogenate absorbance spectrum and standard curve.		
Spectrophotometry. Using a standard curve to assay beet tissue exposed to acetone and methanol.		
Enzymes. Acid phosphatase assays of various plant tissues.		
Mitosis and meiosis. Microscope observations of cells dividing, and paper diagrams.		
Mendelian Genetics. Corn monohybrid and dihybrid crosses, Chi square tests, problems.		
DNA fingerprinting and forensics. Restriction enzyme cuts of DNA, and gel electrophoresis of DNA fragments.		
Plasmid DNA transformation of bacteria, agar plating, Glo gene expression.		

6. If we were to keep the overall best 3 labs, which 3 labs would you recommend we keep, and why? Provide a brief explanation.
7. If we were to eliminate 3 of these labs, which 3 labs would you recommend we eliminate and why? Provide a brief explanation.

- Which labs helped you to learn the most biology facts and concepts?
- Which labs helped you to learn the most lab skills?
- Which labs were most interesting to you?
- Which labs had the most difficult concepts to understand?

- Which labs required the most time to complete, when including time spent outside of lab, either to understand it or to write it up?

The student's rankings of the labs by these questions formed the outcomes assessment. It revealed the student's ideas of their development in skills, content, and inquiry. The results were compiled with a frequency table (Table 6.1), counting which labs students felt helped them learn the most, were most interesting, were most difficult, or were most time-consuming.

Table 6.1. Average Rankings of Labs on the Biology Lab Survey

(Rank of 1 is most and 10 is least. *N* = 91 students)

	Average rankings from 1–10 (1 is most, 10 is least)				
Labs	**Q1. Learn facts concepts**	**Q2. Learn lab skills**	**Q3. Interesting**	**Q4. Difficult**	**Q5. Most time**
Intro to scientific method	7.4	6.9	8.1	8.3	8.3
Microscopy	6.1	4.6	6.5	8.2	7.7
Water relations, osmosis	5.3	5.9	5.7	5.9	5.4
Spectrometry, standard curve	5.1	4.5	5.4	5.3	4.7
Spectrometry, assay tissue	5.8	5.2	5.8	5.2	4.8
Enzyme assays	5.3	5.9	5.3	4.4	4.7
Mitosis and meiosis	4.1	6.4	5.7	5.8	5.9
Mendelian genetics	4.7	6.4	5.0	5.9	5.3
DNA fingerprinting	3.8	3.5	2.8	5.3	4.1
DNA transformation, bacteria	5.0	4.5	3.9	5.1	5.9

A second survey was also given to the students at the same time, the biology attitude scale (Russell and Hollander 1975). This survey measures how comfortable students are with biology (Figure 6.2, p. 80). This survey consists of a set of statements, each expressing a feeling toward biology, such as, *I am always under a terrible strain in a biology class* or, *Biology is fascinating and fun*. Students were asked to strongly agree, agree, be undecided, disagree, or strongly disagree. The results were quantified by assigning numbers to each response (strongly agree = 1, agree = 2, undecided = 3, and so on) and compiled with a frequency table (Table 6.2, p. 81).

Figure 6.2. The Biology Attitude Scale Given to Students at the Same Time as the Lab Survey

Name ___________________________

Each of the statements below expresses a feeling toward biology. Please rate each statement on the extent to which you agree. There are no right or wrong answers. Your response will have no impact on your grade. Thank you for responding thoughtfully.

Place an "X" in the cell of your response to each statement. For each, you may:

	Strongly agree	Agree	Be undecided	Disagree	Strongly disagree
1. Biology is very interesting to me.					
2. I don't like biology, and it scares me to have to take it.					
3. I am always under a terrible strain in a biology class.					
4. Biology is fascinating and fun.					
5. Biology makes me feel secure, and at the same time is stimulating.					
6. Biology makes me feel uncomfortable, restless, irritable, and impatient.					
7. In general, I have a good feeling toward biology.					
8. When I hear the word "biology," I have a feeling of dislike.					
9. I approach biology with a feeling of hesitation.					
10. I really like biology.					
11. I have always enjoyed studying biology in school.					
12. It makes me nervous to even think about doing a biology experiment.					
13. I feel at ease in biology and like it very much.					
14. I feel a definite positive reaction to biology; it's enjoyable.					

Table 6.2. Frequency Distribution Table of Raw Results on the Biology Attitude Scale

(For example, 29 students strongly agreed with Q1 that biology is interesting. N = 91 students)

	Strongly agree	Agree	Be undecided	Disagree	Strongly Disagree	Average score
	#1s	#2s	#3s	#4s	#5s	
Q1. Biology interesting	29	45	9	6	2	2.0
Q2. Don't like; scares me	1	7	7	48	28	4.0
Q3. Terrible strain	6	17	15	37	16	3.4
Q4. Fascinating and fun	13	50	18	10	0	2.3
Q5. Secure; stimulating	4	27	38	21	1	2.9
Q6. Restless; irritable	0	10	20	38	23	3.8
Q7. Good feeling	16	50	13	11	1	2.2
Q8. Hear word; dislike	1	10	6	51	23	3.9
Q9. Approach hesitation	3	14	10	49	15	3.6
Q10. Really like bio	17	44	17	11	2	2.3
Q11. Enjoyed studying	25	36	7	20	3	2.3
Q12. Nervous do expt	0	6	6	53	26	4.1
Q13. At ease in bio	10	40	23	17	1	2.5
Q14. Definite positive	16	38	20	17	0	2.4

The negatively framed questions were converted so a lower value is a more positive response. Specifically, the new value for these questions equals 6 minus the original value so 1 becomes a 5, 2 becomes a 4, and so on. The questions converted in this way were questions 2, 3, 6, 8, 9, 12. After converting these questions, a sum of all 14 questions and an average was computed, thus creating a single summary score that quantified how positive each student was about biology. A total of 91 students filled out complete surveys so these individual scores were grouped into percentiles (Table 6.3, p. 82). For example, the most positive 10% of the class (the 90th percentile) scored a sum of 1.36, equivalent to answering most positive (strongly agree) to two-thirds of the attitudinal questions and positive (agree) to the remaining one-third of the questions.

Table 6.3. Percentile Groups of Students Based on the Single Summary Score That Describes How Positive They Were to Statements About Biology From the Biology Attitude Survey

(Negatively framed questions were converted so a lower value is a more positive response. The score is a composite of 14 statements ranked on a scale of 1–5 based on whether students agreed or disagreed with 1 being strongly agree, 3 neutral, and 5 strongly disagree. *N* = 91 students)

Percentile of class based on attitude about biology	Overall score
90th	1.36
80th	1.71
70th	1.93
60th	2.07
50th	2.21
40th	2.36
30th	2.57
20th	2.79
10th	3.14

Results From Attitude Scale

The results show that students' overall attitude about biology was very positive (Table 6.3). A total of 80% of the class scored 1.0 to 2.79 indicating that they are on the positive side of undecided. Even more, the top 35% of the class scored values from 1.0 to 2.0 demonstrating that they had strongly positive to positive attitudes about biology in the survey. Only the bottom 10% of the class was slightly negative in their attitudes given a 10th percentile value of 3.14. As a consequence student responses were sorted into two groups based on their attitudes. Those most positive about biology, the 35% of the class with average scores from 1.0 to 2.0, were compared with those less positive about biology with average scores below 2.0 in *t*-tests of ranks of particular labs from the biology lab survey. Table 6.4 (p. 87) contains the average rankings of the labs for these two groups of students.

Results From the Biology Lab Survey

DNA Labs

Ranking student responses to questions 1–3 about learning the most facts, concepts, lab skills, and being the most interesting, demonstrated a clear strong preference (Table 6.1, p. 79). The DNA fingerprinting lab was the highest ranked lab with a score of 3.8, 3.5, and 2.8 for questions 1–3, respectively. This score was statistically higher than the score of the next highest lab ($p<0.001$). A total of 41 out of 91 students surveyed ranked it as the most interesting lab all semester and 22 students ranked it number 1 in learning the most biology facts and concepts. The second most preferred lab was the plasmid DNA transformation of bacteria lab, with ranks of 5.0, 4.5, and 3.9 for questions 1–3, respectively. This score was higher than the score of the next ranked lab, and the difference was significant ($p=0.05$). Together these two labs had more first- and second-place rankings than any other labs all semester. These rankings were not affected by student's attitudes about biology.

These labs were complex and required understanding advanced concepts and integrating genetics, molecular biology, biochemistry, and new hands-on skills to carry them out. Students spent five weeks on these labs, and they were the most expensive and time-consuming to supply. A goal of these labs was to have students doing DNA manipulations using procedures and equipment current in research and forensic laboratories. As faculty, we consider the skills students learn in these DNA labs to be relevant and reinforce difficult core concepts with hands-on learning.

The content of the DNA fingerprinting and forensics lab followed standard research protocols for cutting DNA with EcoR1 restriction enzyme and separating the fragments based on size on an agarose gel with electrophoresis. The DNA was sourced from a simulated crime scene or paternity analysis. A student gel is shown in Figure 6.3 (p. 84), from which it is apparent that suspect 3 has a DNA match with the crime scene DNA. Students gave oral courtroom-style presentations of their results and were challenged to provide explanations at a level understandable by a jury. The forensic component was interesting and motivating to students with a wide range of interests.

The plasmid DNA transformation of bacteria lab introduced a standard research protocol used to move genes from one organism to another. The plasmid contains genes of interest that can be moved into bacteria, which then incorporate the new genes and code for new traits. The plasmid contains a gene for resistance to the antibiotic ampicillin. Ampicillin can be selectively incorporated into the growth medium for the bacteria (or not), allowing bacteria to grow or not, depending on whether they have been able to acquire new traits from the plasmid during transformation.

Figure 6.3. Gel of Student Results From DNA Fingerprinting Lab

(Samples of DNA cut with restriction enzyme run on an agarose gel to sort by size. The smallest DNA fragments run from the top well to the bottom of the gel. It is apparent that the banding pattern of the crime scene DNA closely matches that of suspect 3. Lane 1 is standard size markers. Lane 2 is crime scene DNA. Lane 3 is DNA from suspect 1. Lane 4 is DNA from suspect 2. Lane 5 is DNA from suspect 3. Lane 6 is DNA from suspect 4. Lane 7 is DNA from suspect 5.)

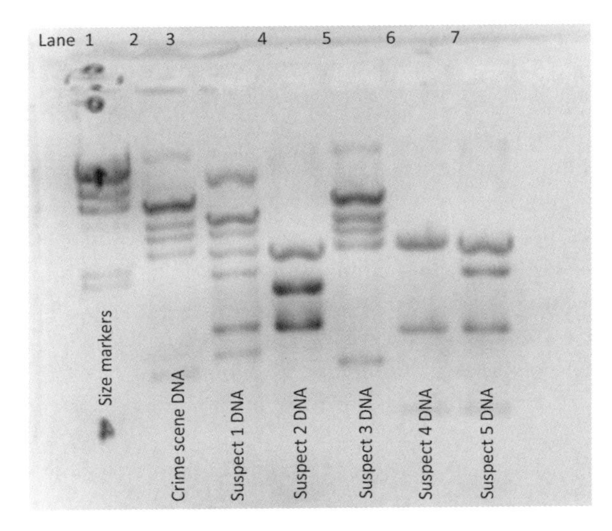

The plasmid also contains a gene originally obtained from jellyfish that codes for green fluorescent protein (GFP), but the protein is only expressed if the sugar arabinose is added to the nutrient medium. The protein causes the bacteria to glow a brilliant green color under ultraviolet light, confirming that the GFP gene is active, and identifying which colonies contain the transformed bacteria. Figure 6.4 illustrates how easily visible these new traits are when the bacteria are grown on different mediums and exposed to ultraviolet light.

Figure 6.4. Example of Results From Plasmid DNA Transformation of Bacteria Lab

(Bacteria were either transformed with plasmid DNA containing an ampicillin antibiotic resistance gene and the GFP gene, or not (the control). The GFP gene codes for Green Fluorescent Protein, which is expressed (synthesized) when transformed cells are grown in nutrient medium (LB agar) with added arabinose, a sugar. The GFP causes cells containing it (here bacteria) to glow a brilliant green color when exposed to ultraviolet light. The green glowing bacterial colonies with UV light in Plate 5 are the descendents of the bacteria that were transformed with the plasmid DNA containing the GFP gene. In this manner genes of interest can be moved into bacteria and detected if the transformation was successful. Plate 1, a control, shows a continuous lawn of bacterial colonies since none have been transformed, and the nutrient medium supports all growth. Plate 2, a control, shows no bacterial colonies since none have been transformed, and with the nutrient medium containing the added antibiotic ampicillin, none can grow. Plates 3 and 4 show only about 10 bacterial colonies that can grow with the added ampicillin antibiotic, and these are the bacteria successfully transformed with the plasmid DNA. Plate 5 shows the green glowing bacterial colonies with UV light successfully expressing the GFP gene when the nutrient medium contains the sugar arabinose. These are the transformed bacteria of interest. Plate 6 shows the transformed bacterial colonies not glowing in UV light since the medium does not contain arabinose.)

Nutrient medium. No GFP DNA. No amp resis DNA.	Nutrient medium with ampicillin. No GFP DNA. No amp resis DNA.	Nutrient medium with ampicillin and arabinose. With GFP DNA. With amp resis DNA.	Nutrient medium with ampicillin. With GFP DNA. With amp resis DNA.

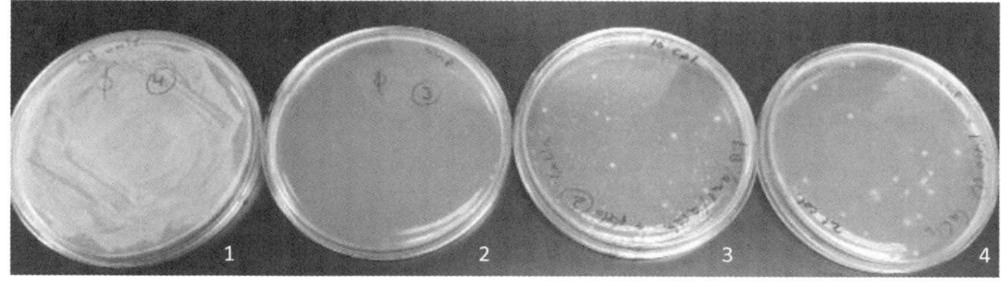

See lawn of colonies	See no colonies	See ~10 colonies	See ~10 colonies

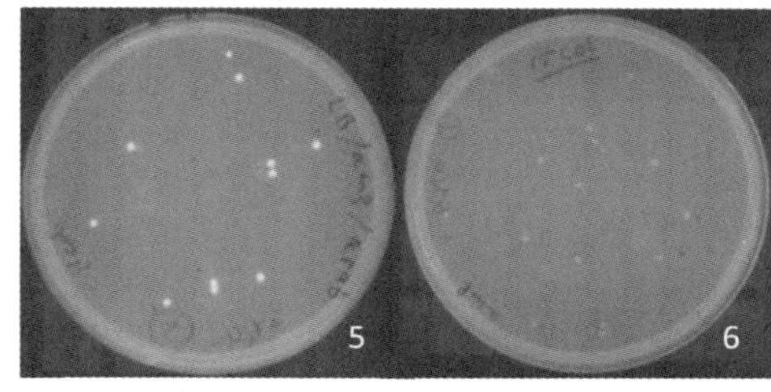

See colonies are green with uv light	See colonies are NOT green with uv light

The contrast between the expression of the ampicillin resistance gene and the GFP gene provides venues for students to understand complex genetic mechanisms. The important value of these techniques is that they provide an accessible and very real illustration of gene expression, mechanisms of genetic selection, and ways that organisms acquire new traits. Students see that they can further develop and apply these techniques for independent learning and research.

Student comments reveal that it was interesting because it was real. Much of the equipment had been borrowed from a research lab, such as pipetters, gel boxes, camera, an ultraviolet light, and petri plates. One student said she loved all the fancy equipment, it felt "sciency." Another student wrote, "They are the most interesting to me and they are really fun to do." It was motivating because it worked well, and students could see it working. When the gel was running there were bubbles rising from the electrodes, and the blue dye front was advancing. Complex ideas that were talked about in lecture were explained again and reinforced by the procedures. Bain (2004, p. 31) states, "People learn best when they ask an important question that they care about answering, or adopt a goal that they want to reach."

Students did not prefer these labs because they were easy or quick; on the contrary, they ranked these labs among the most difficult to understand and time consuming (Table 6.1). The DNA fingerprinting lab was ranked 5.3 and 4.1 for questions 4 and 5, and for the same questions, the plasmid DNA transformation lab was ranked 5.1 and 5.9. A challenging aspect for teachers is that this set of techniques, Restriction Fragment Length Polymorphism (RFLP) analysis due to Single Nucleotide Polymorphisms (SNPs), has been dropped from the current version of the textbook, *Campbell's Biology 9th edition*, although it was in the 8th edition (p. 417).

Spectrophotometry and Enzyme Labs

Other labs that students ranked highly for learning facts, concepts, lab skills, and being interesting were a set of three labs designed to teach spectrophotometry skills and enzyme assays (Table 6.1, p. 79). These rankings of high learning were even higher for the 35% of students who had a higher biology attitude summative score (Table 6.4). T-tests for differences in sample means of ranks were significant for 2 of the 3 labs, 4.7 to 5.2 $p=0.13$ for spec1, 5.1 to 5.8 $p=0.05$ for spec2, and 5.1 to 5.8 $p=0.05$ for the enzyme lab. This survey statistic means that students who strongly agree with positive statements about biology indicated they learned even more, and found it even more interesting, than students who simply agreed. It is also notable that students were not put off by the difficulty of the labs: across the 10 labs for the semester these 3 labs were ranked as 3 of the 4 most difficult ($p<0.001$). Reassuringly, these results show that students have robustness for learning even if the content is difficult.

These labs were complex and required extensive wet chemistry, instrumentation, calculation, timing, and computer graphing to complete. During this three-week sequence the skills and content of each lab reinforced and continued from the previous week. They were inexpensive and relatively simple to supply.

Table 6.4. Average Rankings of Labs on the Biology Lab Survey

(Combining questions 1, 2, and 3, for an average, and combining questions 4 and 5 for an average. Results are presented for the 35% of students most positive about biology (overall score in Table 3 less than 2.0), and for the 65% less positive, from the Biology Attitude Survey. *N* = 91 students)

	Average rankings from 1–10 (1 is most, 10 is least)			
	From students with *more* positive biology attitude		**From students with *less* positive biology attitude**	
Labs	**Q1, Q2, Q3. Learn facts, concepts, skills; interesting**	**Q4, Q5. Difficult; Time-consuming**	**Q1, Q2, Q3. Learn facts, concepts, skills; interesting**	**Q4, Q5. Difficult; Time-consuming**
Intro to scientific method	7.5	8.2	7.5	8.4
Microscopy	5.7	8.1	5.7	7.9
Water relations, osmosis	5.6	5.9	5.7	5.4
Spectrometry, standard curve	4.7	5.1	5.2	4.8
Spectrometry, assay tissue	5.1	5.0	5.8	4.9
Enzyme assays	5.1	4.9	5.8	4.2
Mitosis and meiosis	5.7	6.7	5.2	5.3
Mendelian genetics	5.3	5.7	5.3	5.5
DNA fingerprinting	3.1	4.8	3.5	4.7
DNA transformation, bacteria	4.0	5.4	4.8	5.7

The labs began with living respiring plant tissue, a fresh beet. A measured quantity of tissue was homogenized and first used to determine an absorbance spectrum and the peak absorption wavelength. Measured quantities of water were used to produce a standard curve in units of grams beet tissue per milliliter water. The second week, beet tissue cores were exposed to treatments that would disrupt the cell membranes. These treatments were organic solvents, such as acetone, methanol, and ethanol, and in another experiment, temperature extremes. The standard curve constructed the first week was used the second week to quantify the grams of tissue disrupted. The third week continued concepts and skills from the previous weeks to assay the amount of acid phosphatase, an enzyme that hydrolyzes a phosphate group from many substrates in many types of cells. The activity of this enzyme was used as an indicator of the effects on the biochemistry of the cell of various treatments. The specific treatments were decided by students as they generated and tested hypotheses about the effect of various environmental exposures on living cells. Students chose a wide variety of cell types, mostly various vegetables, and interpreted and communicated their results in written and oral discussion.

Students wrote on the survey that these labs were successful because "they are the most realistic labs for future biology labs" and the labs "all are challenging biology topics that I will use in my career." Students were motivated to generate data themselves. They had to keep track of quantities, use calculations, draw graphs, and then use the graph to convert information (standard curve). They conceived of their own hypothesis, did their own experiments, negotiated the equipment, and analyzed the results. All these activities were helpful to students, as one wrote, "good explanation of concentration." Another stated, "illustrates a great concept in a clear lab." One student said, "by that time, it wasn't so stressful trying to figure out how to work the spectrophotometer." Another, "all of these labs really got the mind engaged and weren't that hard to do at this level." Students quickly learn their answers are neither right nor wrong, but must derive from their data.

Generating and Testing Hypotheses Lab

The calculation of average rank of student responses to the lab survey revealed that one lab scored significantly below all the other labs (Table 6.1, p. 79). Students indicated that they learned the least facts, concepts, skills, and were least interested in the lab that introduced the scientific method ($p<0.001$ when compared to the next low lab). Students also ranked this lab as the easiest and quickest, questions 4 and 5 on the survey, but it was not significantly easier and quicker than the next low lab ($p=0.17$). Student attitudes about biology did not affect the relative ranking of this lab (Table 6.4, p. 87; $p=0.48$, t-test for difference in sample means). This is consistent with earlier results that students do not necessarily find easy, quick labs interesting, or learn much from them.

This introductory lab consisted of a narrative that explained the scientific method, hypothesis testing, experimental design, and valid interpretations of data. It also included several data tables that students were asked to consider, generate an appropriate hypothesis, and graph the data in a way that highlights the distinctions of the hypothesis.

Student comments reflected the opinion that labs should closely follow lecture content, "I recommend that [other labs instead] be kept because they all contain a large amount of useful information and they provide good resources for lecture exam preparation for when the textbook material can seem overwhelming." This outlook was shared by another student, "not very helpful, confusing." One found it worthwhile "because of the experience of lab methods."

Perhaps this lab is weak because starting inquiry is rough. Inquiry viewed as developing scientific thinking, scientific understanding, and science as a hypothesis-driven process is difficult to initiate. Students need time to adjust to thinking in lab and trusting that the "correct answer" is not a single answer but a conclusion consistent with what the data reveal.

Next Steps

Student responses to surveys on their labs are clear: the most interesting labs for students to learn concepts and skills can be difficult and time-consuming, but they must be real, and have relevant challenges. Students had very positive attitudes about biology in general. With this information, our next steps are also clear: continue to develop engaging labs to build core competencies. Then we will motivate students to be the active agents of their own education and unleash the greatest energy source available on campus.

Acknowledgments

I thank IUP student Kaitlyn Robson for technical assistance and constructive suggestions to clarify the student perspective. I thank Cuong Q. Diep for valuable input and for providing excellent plasmid DNA transformation photos.

References

Bain, K. 2004. *What the best college teachers do*. Cambridge, MA: Harvard University Press.

Brewer, C. A., and D. Smith. 2009. *Vision and change in undergraduate biology education: A call for action*. Washington, DC: American Association for the Advancement of Science.

Campbell, N. A., and J. B. Reece. 2008. *Biology*. 8th ed. San Francisco: Pearson Benjamin Cummings.

McManus, D. A. 2005. *Leaving the lectern: Cooperative learning and the critical first days of students working in groups*. Bolton, MA: Anker Publishing.

National Research Council (NRC). 1996. *National science education standards*. Washington, DC: National Academies Press.

Palmer, P. J. 1998. *The courage to teach: Exploring the inner landscape of a teacher's life*. San Francisco: Jossey-Bass.

Russell, J., and S. Hollander. 1975. A biology attitude scale. *The American Biology Teacher* 37 (5): 270–273.

Wilson, O. M. 1967. Teach me, and I will hold my tongue. In *Improving college teaching,* ed. C. B. T. Lee, 7–35. Washington, DC: American Council on Education.

Assessing Learning Outcomes of the Case Study Teaching Method

Linda L. Tichenor
University of Arkansas—Fort Smith

Setting

Arkansas leads the nation in the percentage of students enrolled in small, rural schools; the number of children living in poverty in these areas is among the highest in the United States. Of the public schools, 42% are in rural areas. Based on results of NAEP tests, proficiency in math and science remain relatively low.

Although 97% of public secondary school teachers hold certificates in their main teaching assignment, Arkansas continues to focus on preparing teachers to be better classroom instructors. Westark Community College in Fort Smith joined the University of Arkansas system in 2002 and began offering new programs in K–12 teacher education. This growth into university status has created a rapidly growing number of new and innovative programs designed to raise the educational achievement level of residents of the western Arkansas service area (including rural Eastern Oklahoma) to meet or exceed the national averages. The institution is committed to increasing student learning, first and foremost, over all other priorities with hopes of strengthening the educational, cultural, and economic development in area communities.

The newly emerging disciplines within the College of Science, Technology, Engineering, and Mathematics (STEM) designed baccalaureate programs in collaboration with the College of Education K–12 teaching programs. The unusual aspect of these developing programs was that the two colleges collaborated in embedding the educational program within the individual discipline programs such that the Biology B.S. degree program has a Teachers' Licensure option to prepare high school biology teachers. The chemistry and mathematics departments developed similar programs. Through these collaborations, strong content programs were established. The university hired several new faculty members who were qualified to teach within the discipline and in the education college as well. As a specialist in college science teaching, I was hired to work in the newly developing programs in biology.

That being said, I submit that current models, now firmly embedded in undergraduate science instruction are, unfortunately, not working to improve the plight of college science teaching nor K–12 science education. If we wish to improve science learning and teaching for everyone concerned, we must use models and recommendations such as the National Science Educational Standards as we think about our college level science courses. I was assigned to teach Introductory Biology, which was a course for majors, majors with teacher's licensure, and nonmajors alike. (We no longer have one course for everyone but separate courses for majors versus nonmajors.) I wanted to model best practices for the students involved in the K–12 education programs. This chapter describes the implementation of an introductory biology course using case studies as the organizing principle, not textbook chapters.

Student learning outcomes were assessed using indicators other than grades on multiple-choice type examinations (graded using Scantron™ forms). Learning objectives were broadened to include improvement in thinking skill, writing skills, independent learning, and dispositions toward science; therefore, different assessments had to be designed to determine student success. Grades were based on achievements in thinking analytically, working independently, and writing.

Remodeled Course Using Case Studies

The use of case studies has been described for many years in science education, primarily medical sciences (Broder et al. 2001; Hutchings 1993; Herreid 1994), but what is needed are reports on *how* instructors are using cases in their classrooms and what the students are actually learning as a result.

I was fortunate to attend one of the SUNY–Buffalo Case Study workshops during the summer of 1999 (Clyde Freeman Herreid, Director). According to Herreid (1994), the use of case studies holds promise in science courses for promoting critical thinking but has had "little trial among teachers." It occurred to me during the workshop, why not try to organize an entire course around problems presented in case studies? I wanted to know if the case study teaching method could address current issues in science education, such as increasing students' positive scientific attitudes, developing better cognitive skills, and promoting deep understanding of science. Could I teach an entire course by embedding the content within the case studies? Could I move away from content-as-organizer to a problem-based learning model? Moreover, could I then evaluate the efficacy of the teaching method?

Although there are reports of case study use in teaching science, assessment of course efficacy has lagged behind to date. Hobson (2001) described a physics course in which he attempted to make science more relevant to students' lives for the sake of scientific literacy. In the assessment of course efficacy, Hobson correlated increased class attendance as related to final examination scores and found that those students with attendance records of 85–100% scored 16 percentage points higher than students with an attendance record of 0–50%. In this chapter, I will describe the implementation of an introductory biology course using case studies as the organizing principle. I will discuss the student learning outcomes of the course delivery method using additional indicators to examine evidence of student learning in nontraditionally taught courses.

Introductory Biology was a one-semester, freshman-level course taught in the traditional format of 50-minute lecture periods three times a week plus a two-hour laboratory once a week. The content sequence follows major biological concepts as outlined by a standard biology textbook, and the delivery of the course had not been altered for many years, with the primary teaching methodology as a "stand-and-deliver" model.

I decided to alter my section of the course to reflect National Science Education Standards, which recommend more emphasis on student interests, inquiry process, opportunities for discussion and debate, and shared responsibility for learning. I hoped to get students excited about their own learning and to change their perceptions about scientific work. The best way to accomplish this is to engage students in relevant biological questions. I especially wanted any education majors to experience a different type of class in science. (I will speak to this issue later in the Lessons Learned section.)

Other goals for the course included increasing students' writing and reasoning skills, helping students to become independent of the instructor in pursuing information; and having them learn how to conduct research in areas that interested them most about material in the case studies. I wanted them to understand content, but more on a need-to-know basis rather than because it was the next chapter in the book.

I had previously written a supplement of 30 case studies to accompany Dr. George B. Johnson's, *The Living World*, 2nd ed. (2000), a biology textbook published by McGraw-Hill. The website where the cases may be accessed is *http://auth.mhhe.com/biosci/genbio/casestudies*. The topics in these case studies were written to be relevant to students. Previously, I had used these cases sporadically to supplement more traditionally taught material. What I really wanted to try was designing an entire course using the problems raised in the case studies to serve as the scaffolding upon which to build the sequence not the chapter number. Readings in the textbook would be necessary for understanding background information needed to solve the cases. Outside research was a "must," because questions asked in the case studies could not be answered solely by reading the content in the textbook.

In order to reorganize the course around the case studies, I arranged 12 cases into 8 learning units whose topics corresponded to the traditional syllabus used by other faculty members teaching the course. Each learning unit used one or two case studies from the website mentioned above.

Major Features of the Instructional Program—One Example of a Learning Unit

Overview

The unit on energy conversion (photosynthesis and respiration) is always difficult for students. Metabolic pathways seem esoteric, especially if the students are nonmajors and have not had any chemistry or biology preparation in high school. It is difficult for an instructor to maintain students' interest in learning if they find the material irrelevant. However, by using the drama of human lives in danger, as in the case study entitled *Averting Disaster in Biosphere 2* (Figure 7.1, p. 95), I was able to make the major points of photosynthesis and respiration seem real and exciting.

The plot of the case study is as follows: Human beings were enclosed in the Biosphere 2, a super-size greenhouse in Oracle, Arizona, for two years to test if life could be sustained in a closed environmental system. The "biospherians" were doing just fine until a type of soil microbe began to overproduce and give off more carbon dioxide than could be used by the green plants in the closed system. The increased levels of carbon dioxide became toxic to the humans, producing severe respiratory acidosis, a condition requiring hospitalization. What better way is there to look at photosynthesis and respiration than in a dramatic life and death situation?

The assignments for the unit were to address the questions in the Need to Know section and to complete one of the two suggested assignments in the Assignments section (see Figure 7.1, Case Study). We could master a case in about a week of classroom work. Evaluations of the writing assignments were based on preplanned rubrics and were good indications of what students actually learned about the material.

Classroom Management

Students were expected to read the case over before class. I would then present the case to the students during class and emphasize the relevance and importance of the issues. The cases were divided into smaller parts that had questions posed as "What We Need to Know." Students were expected to address the concepts by reading the textbook or doing online research. The in-class work involved the students working in groups of four to five to answer these Need to Know questions. Often, I addressed students' questions by circulating around the room and working directly with the groups. If several groups had the same question, I would go to the front and explain to all. A portion of the classroom activity included a brief, minilecture describing difficult concepts. It would take two or three class periods to get the answers to these questions researched and answered. Students had lively conversations and would sometimes become distracted by extraneous "research." Part of my job was to keep the students on-task as I circulated around the room.

When all questions about the case were answered, I assigned a homework paper derived from the Assignment section of the case study. Students had the option to choose among questions. I designed questions that would appeal to different interests and majors.

If, for example, a student were a science major, he or she may wish to write a paper on the first assignment choice (see Figure 7.1, Assignments). The assignment instructs the following: *Design another solution to remove the* CO_2 *from the atmosphere in the Biosphere. Write up a two-page proposal to persuade the granting agencies to give you money to carry out your ideas.* On the other hand, a business major taking my class may select the following assignment: *Write a position paper to encourage automobile manufacturers to increase the gasoline mileage on all new cars produced after 2004 based on your knowledge about the events in the Biosphere 2.*

Results of Assessment of Learning Outcomes

In assessing the efficacy of the case study method, I wanted to evaluate the expanded set of objectives described previously by using qualitative assessment of data as well as final course grades to represent student achievement. The purpose of qualitative research in this case was to produce findings about course effectiveness and not necessarily to contribute to learning theory (Patton 1990). I wanted to generate information about the overall effectiveness of the case study method as used in one introductory course and needed to know how students perceived the learning

Figure 7.1. Case Study

Averting Disaster In Biosphere 2

Julia Brink (fictitious character) looked forward to her two-year commitment to living in the experimental ecosystem called Biosphere 2 (actual location). The Biosphere 2 is located about 40 miles from Tucson in a small Arizona town called Oracle. As she drove up the winding road to the Biosphere 2, she was filled with anticipation. She looked forward to meeting her habitat partners. The three-acre facility looked like a giant greenhouse looming in the distance. She knew that the seven other team members shared her aspirations of being one of the first humans to live in a controlled environment that mimicked what it might be like to live in outer space during long journeys to distant planets. In a sense they were inner space astronauts!

Julia wondered, "Could it be possible to live inside a glass-covered structure for two years without coming out?" The Biosphere 2 was an experiment as a "closed ecosystem," much the same as an enclosed terrarium. A terrarium contains plants that produce oxygen and use CO_2. The ecosystem includes animals as well as bacteria and fungi. The air in the Biosphere 2 is recirculated. Outside air never enters the enclosed space. The Biosphere 2 relies upon the plants to produce enough oxygen and remove CO_2 for the human and the other animal inhabitants.

Inside the biosphere are several "biomes," including a human habitat, an ocean, a tropical rain forest, a savanna, a marsh, and a desert. The humans lived off the land, growing their own food, even their own coffee beans!

Part I—What We Need to Know

1. Briefly describe the Biosphere 2 as a closed ecosystem. What is a "closed" ecosystem?
2. Review the processes of photosynthesis and cellular respiration. **(Lecture)**

   Do both plants and animals produce CO_2? Do both plants and animals use oxygen? Do both plants and animals use CO_2? Discuss these ideas.
3. Where do the animals enclosed in the giant greenhouse get their oxygen supply? Why doesn't CO_2 build up in the biosphere atmosphere?
4. Discuss the first and second law of thermodynamics. **(Lecture)**

   Do these laws apply in the biosphere? Why or why not?
5. Describe the general characteristics of the various "biomes."

After about one year of living in the Biosphere 2, a sudden emergency arose. Even though there were plenty of plants under the dome, the carbon dioxide levels had been slowly rising to the point of causing the humans inside to suffer from a medical condition known as respiratory acidosis. That is a condition which occurs when CO_2 builds up in the bloodstream to a level too toxic to sustain life. The situation was grave. Would they have to abandon their mission?

Part II—What We Need to Know

1. What happens to animals when CO_2 builds up in their environment? **(Lecture)**
2. How much CO_2 is there in the outside atmosphere?
3. Predict the reason for the CO_2 levels increasing in Biosphere 2.

Julia called together her colleagues to problem-solve the situation before they had to abort the mission early. Was there anything that they could do to stop the rising CO_2 level? Julia believed that if there were some method of removing CO_2 from the circulating air, they could remain in the Biosphere 2 indefinitely.

The solution designed by the team was to recirculate the "used" air over and over again through a large water system containing algal mats. The mats were located in a room about the size of a football field. The mats were covered with common river algae. Flowing over the mats was a constant stream of water aerated with the recycled air. Low and behold, the solution proved to work! In time, the carbon dioxide levels returned to normal, and the mission was saved.

Part III—What We Need to Know

1. Why would soil microbes unexpectedly begin to overpopulate Biosphere 2?
2. What type of metabolic process would produce large amounts of CO_2 in soil microbes? **(Lecture)**
3. Why would algal mats serve to remove CO_2 from the recycled air?
4. How would the CO_2 from the recycled air get into the water that recirculated over the algal mats?
5. Why did the team not simply punch a hole in the Biosphere 2 to allow the CO_2 to escape?
6. What was the actual outcome of the Biosphere 2 experiment?

Part IV—Assignments

1. Design another solution to remove the CO_2 from the atmosphere in the Biosphere. Write up a two-page proposal to persuade the granting agencies to give you money to carry out your ideas.

2. Write a position paper to encourage automobile manufacturers to increase the gasoline mileage on all new cars produced after 2004 based on your knowledge of the events in the Biosphere 2.

experience. I wanted to discover the major strengths and weaknesses of the methodology in order to make the course more effective for future course design, and finally, make recommendations for any instructor wishing to teach by using case studies. Therefore, I designed an open-ended questionnaire that addressed the learning experience from the students' perspective and administered it during the last week of the course. A copy of the instrument is demonstrated in Appendix 7.1, p. 105.

I summarized the results of the questionnaire by forming analyst-constructed typologies, which are recurring statements in the data, and sorted these into categories for each question. I quantified the typologies to demonstrate trends of students' experiences, attitudes, skills, behavior, and perceived changes in content learning.

Seventy-one ($n = 71$) students were asked to fill out the questionnaire, and 62 ($n = 62$) responded (87% respondent rate). The result of question 1, which asked, "Did you *learn* using the Case Study Method of teaching/learning?" was 100% agreement. I was surprised because I have yet to experience total agreement among students in science courses with one teaching method. From past experience, I know that students are honest enough to claim that they did not learn from a course if they perceive the experience is not worthwhile.

To question 2 ("What did you learn about the learning process itself as a result of this method?"), students reported that they learned how to conduct research and work independently as a result of the course (39%). Others reported that they actually preferred researching on the internet and reading other sources to reading a textbook (13%). They reported that the course was actually more interesting than they had anticipated (24%) because of association of the cases with real-life situations. The following are three students' responses. The words in parentheses are mine.

- You do not have to sit around and study for test(s) just to learn about things.
- I learned that I can learn a great deal from just researching topics on my own as opposed to listening to class lectures.
- I learned that I am a good researcher and very creative. I now know that if I teach myself (and) learn it myself, I am more inclined to remember it than rather just get the answer from the instructor. I retain it more and am able to use it in other classes that I am currently taking.

Question 3 asked students to reveal what they learned about *themselves* as learners as a result of this teaching method. Fifty eight percent (58%) of the students reported that they actually enjoyed researching material, that they learned better by researching than by reading textbooks, that they would more likely retain the information better, and that they learned self-discipline. They also reported that they liked science more than they thought, learned to write more efficiently, learned how to relate science with real-life experiences, and expressed interest in using the case study method eventually in their own teaching career. Comments included:

- This is the best way for me to learn. I do not test well at all. I retained most of what I learned.
- Personally, I learned that my study habits are not where they need to be. I have a tendency to procrastinate and have to rush to get everything done, and sometimes I did not get the assignment done.
- I learned that I can retain much, if not more, of the information I learned through research if did on my own in conjunction with the lectures.
- I learned that if I actually put in the time and effort that I can learn a lot and do a good paper.

Students reported that the major strengths (question #4) of the course were learning to do research outside their textbook reading (44%) and becoming independent learners about a variety of topics. They reported that the information learned was harder to forget than in a typical course because of the real-life problems and open discussions. The notion of retention reappeared in several of the items as well as relationship of science to real life. Comments included:

- It is more appealing to the non-biology major. You're able to get more in-depth with certain areas and get more knowledge about what is taught in the book. This makes the learner aware of how real biology is.
- I was in control of my work. The work was done as I saw fit. I was able to do research on the internet instead of (getting) all the information from a book.
- Makes you think, it's interesting, makes you dig up information that you otherwise would not think about much.
- As I was doing research and then writing the different case studies, I realized I learned more and it stuck with me unlike cramming for a test.

In Question #5, the major weaknesses of the case study method revolved mainly around misunderstanding of assignments and lack of concrete instruction on how to write up cases for the grade. Uncertainty was probably due to first-year students transitioning to a new type of learning method and lack of specificity on my part when outlining the rubric for grading their writing. As the course progressed, they had less trouble producing papers on each case. Since then, I have learned how to better guide students through a writing rubric and understand where they are most likely to find the grading system "fuzzy." Most other comments concerned problems with group work that are commonly reported in collaborative learning. These statements noted that some students did not prepare or participate fully in the group work and that there might be temptation for copying another student's research. Comments included:

- I didn't see any weakness in this method. The class was not set up like "If you don't understand you fail." It is more like "If you don't understand, find out." This could only be a weakness for people who don't care to learn.
- If someone does not understand the assignment, they might not do the work correctly, so therefore they really didn't learn anything about that topic of study.

- It isn't a one night type of assignment. I don't know if that's a weakness or not. I think it is a good way to teach and there aren't many weaknesses.
- Lack of clear directions for some assignments. (I felt I had to guess.) If student lacks a computer or access to computers or the web is down then they cannot retrieve case study or links to additional information.
- It was difficult in learning how to answer the case studies because I didn't feel there was much instruction. It got easier the more I did it though.

In question #6, I asked the students to compare a previous biology course (either high school or college level) with this course. Almost all of the students reported that this course, as compared with past biology courses, was better. They reported that high school biology was "boring" since memorizing definitions, taking notes, filling out worksheets, and taking tests was the typical teaching methodology. Those students who had been unsuccessful in college biology previously and were retaking the course reported that they actually understood biological concepts as never before because of the application of content to real-life problems. Comments included:

- I think I learned more in this class by doing this teaching/learning method than the science classes I took in high school, that (sic) just read straight out of the book and you don't get anything out of reading word for word out of the book.
- This teaching method is much better than the biology I took in the 9th grade. I actually learned about the environment and the things that go on this day and age. I like this method because in the other biology, the teacher just lectured and we took tests.
- Other courses I've had have been stand up and deliver (student learned the terminology from explanations of methodology) and I didn't learn very much from them because I just had to memorize facts. I didn't have to apply what we learned.
- In biology in high school, we went word for word by the book. We had to learn all the little nit-picky stuff. The whole time I was thinking, how is this relevant to me?
- I took another science course during the summer at UAFS and it was completely different from this class. The instructor would read from the book or just lecture (it was his way and words that were correct) and I didn't learn from the course. It's only been seven months and I don't remember anything I learned in there. I like this class because there is much more of learning on your own and finding research that you will retain.
- The only other biology course I had was in high school. And it was really boring compared to this class. We had to memorize every definition and everyone ended up hating that class. But in this class, although there was an emphasis on the book and its content, a greater emphasis was put on learning and researching about current events related to the book work. So then it became more interesting and fun.
- (In other biology courses), I never read my textbook or really even looked at it. In this course I opened my book often and the instructor handed out other sources of information.

When asked for suggestions for improvement (question #7), 55% of the students reported "nothing." Sixteen percent (16%) wanted more details listed in each assignment. I constructed the assignments purposefully open-ended so students could choose their own learning direction as dictated by their own interests and not mine. In the sample case study (Figure 7.1), Part IV directs the students to their assignment and offers them two choices. They could choose either option depending on which seemed most relevant to them.

Students also felt that they needed more time for a single case study. Typically, we covered about one case per week, and in retrospect, I agree with the students. Although they did not use the terms, they were taking issue with the depth versus breadth coverage of the course.

Table 7.1 demonstrates that in traditionally taught sections of introductory biology the percentages of Ds, Fs, and Ws were 51% and 41% respectively for fall semester 2002 and fall semester 2003. However, in the one section taught using case studies, the number of Ds, Fs, and Ws was only 18%. It should be noted that all sections described above were taught by me. The only difference was the lecture format versus case study format. The data indicate that using case studies as described in this study presents less risk of attrition in the non-major biology courses. A grade of "D" was included in the attrition rate because it is considered unsatisfactory.

Table 7.1. Course Grades and Attrition Rates Comparing Traditionally Taught Introductory Biology Sections With a Case Study–Based Course

All sections were taught by same instructor.

Courses Assessed	Letter Grade									
	A	B	C	D	F	W	Total	Number of D,F,W	% D,F,W	% F,W
Fall 02 All Sections	9	32	43	33	41	15	173	89	51%	32%
Fall 03 All Sections	44	84	117	50	61	57	413	168	41%	29%
Case Study Sections	27	28	16	3	4	9	87	16	18%	15%

Summary and Lessons Learned

I have learned many lessons over the years about teaching with case studies. I will innumerate these for others who wish to teach entire courses using ONLY case studies as I did or wish to merely embed cases within their traditionally taught courses.

1. Study the literature that is now available from a multitude of sources on "how to teach using case studies," such as *http://cte.umdnj.edu/active_learning/active_case.cfm*
2. Attend a workshop. The following description comes from the National Center for Case Study Teaching in Science, located at SUNY, Buffalo. *http://sciencecases.lib.buffalo.edu/cs/training*

ONE OF OUR MAJOR AREAS of focus is training science faculty in the case method of teaching. We do this through a number of activities, including an annual summer workshop and fall conference. These are attended by undergraduate, graduate, and high school teachers from all areas of science and technology, including anthropology, astronomy, biology, chemistry, computer science, earth science, engineering, mathematics, medicine, nursing, pharmacy, physics, psychology, science education, and more.

Summer Workshop

Our five-day summer workshop teaches science faculty how to teach with as well as how to write case studies. The first three days provide an overview of the case study method followed by a series of demonstrations in which workshop faculty model a variety of case teaching methods, including the discussion method, interrupted case method, intimate debate, and team learning. The last two days, workshop participants teach a case study they have researched and written during the week to undergraduate students we hire to act as constructive critics.

Fall Conference

Our two-day fall conference covers much of the same material as our summer workshop but in a more condensed format and without the practice teaching sessions. In addition, it includes a second track for more advanced case method teachers, with sessions aimed at helping them to further develop and refine their case teaching skills. In recent years, we have added a third track specifically designed for high school science teachers. The conference also features plenary talks and a poster session.

Training Videos

Our training videos are intended for those of you who cannot attend our summer workshop or fall conference or who simply want to share what you have learned at one of these events with colleagues back home. The videos show real students in real classrooms using cases—in a discussion-based class in one video, and in the context of team learning in the other.

3. Use a mentor with experience in the use of the case study method, whether at your own institution or another. There is nothing worse than wasting time or "reinventing the wheel." Learning how to organize content around the cases requires throwing out some content material and that requires courage. We have been trained that the textbook is a sacred document that we must cover from atoms to ecosystems. I am just now getting used to the idea that you can organize an entire course around the five most relevant concepts in a particular field. You may subject to criticism by colleagues for doing so. A mentor would give you the support to redesign your courses.
4. Engage students in your choice of pedagogy. Thoroughly explain to students the rationale for teaching with case studies and get them interested in a new style of learning.

You will have to make your evaluation procedures extremely clear to avoid negative reactions.

5. Teach yourself something about group work. Group work can become very unpopular if the members of the group do not work together well or consistently. Dr. Bonnie S. Wood describes a method of allowing students to anonymously evaluate their peers in a working group. Her book is *Lecture-Free Teaching: A Learning Partnership Between Science Educators and Their Students* (2009).

In my own experience, colleagues who were teaching other sections of the introductory biology course using the traditional lecture method were skeptical about the effectiveness of the case study teaching method in terms of decreasing the amount of content, students' understanding content without the teacher lecturing, and students' retention of facts. It is clear from colleagues' concerns that, first an agreement must be reached as to *what* (my emphasis) is important for students to learn in our science courses. If learning content is the sole objective of a course, generalizations about learning outcomes can only be based on grades from examination scores. However, in this study, learning objectives were broadened to include improving thinking skills, writing skills, dispositions toward science, and independent learning; therefore, different assessments had to be designed to determine student success. I do not know how many "facts" were actually learned and retained as a result of the case study method, because I did not give typical content examinations, only writing assignments. Students *perceived* that they learned more about biology content when they conducted their own searches because the content seemed relevant and interesting. I based grades on achievements in thinking analytically, working independently, and writing. The results of the questionnaire indicate that students perceived that they, in fact, learned, comprehended, and retained more of content material using the case study method as compared to typical teaching methods.

Overall, the big lesson learned is that instructors can raise learning standards by using techniques that require students to put forth more effort than rote memorization assessed by standard multiple-choice or fill-in-the-blank examinations. Most students in the case study biology course revealed that they spent more time doing research than they would have reading the textbook and reviewing notes for a traditional examination. In this way, inquiry- or problem-based learning, such as the case study method, would seem to raise the quality of education in the sciences, especially if one of the aims is scientific literacy for nonmajors.

Another lesson to be learned from this chapter on the use of case studies to teach science has to do with the changing climate of K–12 education through new standards recently being developed. The *Next Generation Science Standards* (NGSS) were released in April 2013. Currently the basis of state science standards are derived from two documents set forth almost 20 years ago. These are the *National Science Education Standards* (NRC 1996) and *Benchmarks for Science Literacy* (AAAS 1993). A document by NRC, entitled *A Framework for K–12 Science Education: Practices, Crosscutting Concepts, and Core Ideas,* was released in 2012. Padilla and Cooper (2012) discussed the implications of the *Framework* for both K–12 and college science teachers. It is their contention that since K–12 students taught using the new *Framework* will be arriving in college classrooms prepared in a different way than those in our classrooms currently, it would behoove college teachers to be prepared to alter their teaching methods or be perceived to be dinosaurs using the older teaching methods. In their words, "The integration of content and practices means that the current 'mile wide, inch deep' approach to introductory college-level science classes will also have to change" (2012, p. 6). College science teachers will also have to rethink how they teach the K–12 teachers in order to prepare them to teach under the *Framework*. After having reviewed the *Framework*, I cannot help but submit that the use of case-based instruction is one of the types of teaching methods that will accomplish some of the goals outlined, where depth of subject is expected. The statement below is an example:

> [T]he committee is convinced that by building a strong base of core knowledge and competencies, understood in sufficient depth to be used, students will leave school better grounded in scientific knowledge and practices—and with greater interest in further learning in science—than when instruction "covers" multiple disconnected pieces of information that are memorized and soon forgotten once the test is over. (NRC 2012, p. 23)

References

American Association for the Advancement of Science (AAAS). 1993. *Benchmarks for scientific literacy.* New York: Oxford University Press.

Association of American Colleges. 1991. *Liberal learning and the arts and science major: The challenge of connecting learning, volume 1.* Washington, DC: Association of American Colleges.

Broder, J. M., K. Martin, A. Rosenbloom, L. P. Zufan, and H. Klein. 2001. An international survey of case use in higher education: Report of the WACRA case standard setting committee. *www.WACRA.org*

Donald, J. G. 1997. Higher education in Quebec: 1945–1995. In *Higher education in Canada,* ed. G. Jones, 159–186. New York: Garland.

Herreid, C. F. 1994. Case studies in science: A novel method of science education. *Journal of College Science Teaching* 23 (4): 221–229.

Herreid, C. F. 2002. What makes a good case? *http://ublib.buffalo.edu/libraries/projects/casesteaching/good-case.html*

Hobson, A. 2001. Teaching relevant science for scientific literacy: Adding cultural context to the sciences. *Journal of College Science Teaching* 30 (4): 238–243.

Hutchings, P. 1993. *Using cases to improve college teaching: A guide to a more reflective practice.* Washington, DC: American Association for Higher Education.

Johnson, G. B. 2000. *The living world.* 2nd ed. Dubuque: McGraw-Hill.

National Center for Education Statistics (NCES). 1996. *National assessment of educational progress. http://nces.ed.gov/pubsearch/getpubcats.asp?sid=031*

National Research Council (NRC). 1996. *National science education standards.* Washington, DC: National Academies Press.

National Research Council (NRC). 2012. *A framework for K–12 science education: Practices, crosscutting concepts, and core ideas.* Washington, DC: National Academies Press.

Padilla, M., and M. Cooper. 2012. From the Framework to the Next Generation Science Standards: What will it mean for STEM faculty? *Journal of College Science Teaching* 41 (3): 6–7.

Patton, M. Q. 1990. *Qualitative evaluation and research methods.* 2nd ed. London: Sage Publications.

Phillips, C. D., L. L. Tichenor, and R. A. Reese. 2000. Assessing effectiveness of introductory biology laboratory exercises toward increased learning outcomes. Paper presented at the Arkansas Conference on Teaching, Little Rock, Arkansas.

Tichenor, L. L. 2000. Assessment of the learning environmental factors that affect outcomes in a large introductory biology course. Paper presented at the Arkansas Conference on Teaching, Little Rock, Arkansas.

Tichenor, L. L. 2001. Case studies. *http://auth.mhhe.com/biosci/genbio/casestudies*.

Tobias, S. 1990. *They're not dumb, there're different: Stalking the second tier.* Tucson, AZ: Research Corporation.

Wood, B. S. 2009. *Lecture-free teaching: A learning partnership between science educators and their students.* Arlington, VA: NSTA Press.

Appendix 7.1

Student Questionnaire

In your own words...

1. Did you learn from the case study–based method of teaching/learning?

    __________Yes
    __________No

2. What did you learn about the learning process itself as a result of this teaching/learning method?

3. What did you learn about yourself as a result of this teaching/learning method?

4. What was the major strength of this method of teaching/learning?

5. What was the major weakness of this method of teaching/learning?

6. Compare this course teaching/learning method with some other science or biology course that you may have taken in the past.

7. What are your suggestions for improvement on this method of teaching/learning?

Implementing Jigsaw Technique to Enhance Learning in an Upper-Level Cell and Molecular Biology Course

Sandhya N. Baviskar
University of Arkansas–Fort Smith

Setting and Overview

The University of Arkansas–Fort Smith (UAFS) is a public, coeducational, four-year university located in Fort Smith, Arkansas. The historic city of Fort Smith is the second largest city in Arkansas and is located on Arkansas-Oklahoma state border. UAFS is the fifth largest university in the state of Arkansas, with current student enrollment of approximately 7,600. It is one of the 11 campuses that constitute the University of Arkansas System. UAFS is also a part of Arkansas Leadership Academy, comprised of 49 partner institutions, including 15 universities and other professional and government institutions. University of Arkansas–Fort Smith attained state university status from a two-year community college in 2002. It consists of seven colleges that offer bachelor's degrees in arts, sciences, business administration, applied science, nursing, and music education. It also offers associate degrees in arts, applied science, general science and technical certificates and certificates of proficiency.

The students majoring in biology at University of Arkansas–Fort Smith are required to take the cell and molecular biology course (BIOL 4803) during their junior or senior years. This upper-level course is offered only in fall and I have been teaching this course since 2009. The enrollment in this course fluctuates between 25 and 40 students. The course includes a variety of conceptual and current topics in cell and molecular biology. The conceptual topics include protein structure and functions, membrane structure and transport, intravesicular traffic, chemotaxis and cell motility, cell communication and cell cycle regulation. Current topics include cancer, Alzheimer's disease, stem cell biology, apoptosis, gene therapy, genetic modification of food, genome instability, protein degradation, tissue engineering, cellular stress, and aging.

A majority of biology majors find cell and molecular biology very challenging because the topics are loaded with huge volumes of abstract information. When cell and molecular topics are taught using only lecture method and student learning is assessed on multiple-choice tests, students tend to forget the information (Heady 1993). Since students do not get the opportunity to explore and expand the topics further, it not only leads to superficial learning but they are not motivated to pursue graduate programs in cell and molecular biology.

Major Features of the Course

I teach the conceptual topics of the course by interweaving a variety of inquiry-based teaching methods into lectures. These include quizzes, think–pair–share, short videos, demonstrations, and couching relevant questions to the class at appropriate times. The purpose of mixing various teaching methods into lectures is to create a learning environment based on constructivist learning principles. These learning principles are: (1) eliciting prior knowledge, (2) creating awareness of differences between prior knowledge and new knowledge, (3) application of new knowledge with feedback, and (4) reflection on learning (Baviskar, Hartle, and Whitney 2009). I find that delivering a lesson plan based on constructivist learning principles engages students and motivates them to explore the topics further.

The current topics of the course, listed above, are covered using a cooperative learning technique called Jigsaw or "Students teaching other students" (Cooper 1990). Jigsaw is a structured cooperative learning strategy in which a group of students work together to learn about a topic. A topic is divided into four to five coherent segments or subtopics and each student is responsible for one subtopic. Thus, each student contributes and brings unique knowledge and shares his or her expertise with the group (Aronson and Patnoe 1997). The current topics in cell and molecular biology are well suited for the jigsaw learning strategy, as these topics have different facets. For example, the topic "genetic modification of food" can be divided into diverse subtopics such as introduction (controversy, misconceptions), history, and process; benefits and potential risks; and concerns and global status. Moreover, some of these topics are "hot" and controversial, some are of global concern, and some are on the forefront of research; therefore, students can relate to these topics and feel motivated to explore. To become an "expert" on a subtopic and teach it to the other group members, a student is required to research, read, discuss, question, clarify, and write (Myers and Jones 1993). These activities not only foster better retention of subject matter but help expand student's thinking abilities as well (Johnson, Johnson, and Smith 1991; McKeachie et al. 1986). Apart from achieving content goals, jigsaw learning strategy enables participants to enhance skills such as verbal and communication skills, reading comprehension skills, listening skills, and critical-thinking skills (Kohl 1982; Lord 2001). It also increases motivation to learn and promotes a positive attitude toward the subject area (Kagan 1994; Stahle and VanSickle 1992; Sandberg 1995). Working with peers, students enhance their interpersonal skills and develop empathy and respect for each other (Alico 1997). There is good evidence in literature that indicates that peer teaching helps develop poise and confidence by reducing anxiety and evaluation apprehension (Maloof 2004).

Implementation of the Jigsaw Technique

To implement the technique, the class is divided into four-person groups. To divide the work equally in a group, each topic is divided into four segments. For example, the stem cell biology topic is divided into four subtopics: introduction, types of stem cells, research and applications, and political and ethical issues. To assign the subtopics randomly and form heterogeneous groups with regards to gender, ethnicity, race, and ability (Slavin 1995), I bring cards to class with different topics and subtopics printed on them and students are asked to pick one card out of the stack and write their names on it. I collect the cards and make a list of the students with assigned topics and subtopics. The list is then sent to the class as an e-mail attachment so that they know who their group members are. The groups are usually formed in the second week of the semester. After the groups are formed, I give a short lecture to the class about the jigsaw technique and its importance. I also specify objectives and tasks, criteria for evaluation, and desired cooperative behavior that is expected from each member of a group.

To become an expert on a subtopic, a student explores various authentic sources such as scholarly books, peer-reviewed journal articles, and trusted internet websites. The expert student shares his knowledge with his or her group and meets with the instructor to check his or her knowledge and understanding about the topic. At this stage, if the instructor finds that a student lacks important information, then the student is directed to appropriate resources. The group meets several times and shares their findings with each other so that every member of the group becomes knowledgeable about the topic. A group then puts together a PowerPoint presentation on the topic and presents it to the class. The jigsaw presentations are arranged toward the end of the semester (usually last three class periods). Thus, a group usually gets nine weeks for preparation. The group meetings usually occur outside the class in a computer laboratory, library, cafeteria, or lobby. One lecture class period is also reserved for a group meeting.

Jessica Chitwood and Britney Fitzgerald spoke about the concept and history of tissue regeneration to the class (Students' presentations in cell and molecular biology class 2011).

During a 15-minute presentation, each group member speaks about his or her segment and thus entire class becomes knowledgeable about the topic. To encourage class participation and discussion among students about the topic, a presentation is followed by a five to seven minute question-and-answer session. Each group member is required to write three multiple-choice questions from his subtopic, thus a group creates 12 multiple-choice questions on the topic and the instructor distributes printouts of these questions to the class before the presentation begins. After the presentation, the presenters (the group members) orally seek correct answers to the 12 multiple-choice questions from the class. After this, class members ask questions to the presenters, which usually leads to meaningful discussions on the topic.

Michaela Jester and her group enlightened the class about stem cell biology. Michaela explained about different types of stem cells (students' presentations in cell and molecular biology class 2011).

Assessment of Students' Performance

In order to hold every member of a group accountable, each member is assessed individually. For assessment, I have developed a rubric (Table 8.1, p. 116) consisting of five criteria that meet content and noncontent assessment goals. The rubric is sent as an e-mail attachment immediately after the groups are formed. Thus, the rubric serves as a guideline for preparation and becomes an integral part of the learning process. Content knowledge and depth of understanding of a topic or subtopic is assessed from annotated bibliography, quality of multiple-choice questions, and depth and accuracy of the information. Non-content goals such as verbal and communication skills, reading comprehension skills, and writing and computer skills are also assessed, as these goals are embedded in the components of the rubric. A student could earn a maximum of 10, 5, or 2 points on each of the five components of the rubric depending on his or her performance.

Justin Casey and Cole Dalmont talk about the factors that lead to cellular stress. Their group presented on the topic "Cellular Stress and Aging" (students' presentations in cell and molecular biology class 2011).

In fall 2009, I taught the course and implemented the jigsaw technique for the very first time. The design of the technique was not well structured and was not tied to the students' grades. In fall 2010 and 2011 classes, the technique was improved and implemented as described above and tied to the students' grades. Therefore, in this study, the fall 2009 class is considered as a reference or control group.

Evidence of Success

Students' grasp of the course material and their engagement and motivation increased considerably as a result of implementing the jigsaw technique in the cell and molecular biology course. Students' understanding of the material is reflected by their final grades, while motivation and engagement to learn the subject matter is evident from student evaluations and their informal reflective feedback collected at the end of the semester. My senior colleague, who is an expert and proponent of inquiry-based active learning, was invited to critique and evaluate jigsaw presentations in the fall 2011 class. Her comments were favorable but she also suggested further improvements in the implementation of the technique. Here are some of the informal reflective feedback from fall 2009 class (control group) that lead to improvements to the jigsaw technique:

- I like the idea of jigsaw presentations. There are a couple of things I would have wished were different. First, more time to prepare the presentations would have been wonderful. Second, I wish there were some kind of accountability. In my group there were a couple of people who did not really do their part, and one person did not do anything until the last day.
- This project was a lot of fun. I learned a lot about my subject from researching articles, books and websites. It was good to work with other classmates and get their opinions. It

was difficult getting six people get together to work. I would enjoy doing a project like this one in the future. However, I would suggest having smaller groups and breaking each topic into parts.

- I thought that the jigsaw presentations were a great learning experience. The process allowed me to become more involved with the learning process and allowed me to explore sources other than the required textbook. I also believe that giving presentations in groups helps alleviate some of the stress and nervousness of presenting in front of others.
- Putting together the presentation and researching the material really helped me learn the subject but I think it would have been better if I had a semester to prepare.
- I believe the jigsaw presentations were very interesting; however, it was very lengthy and some areas of the presentations lost my attention. Perhaps in the future, the presentations should not focus so much on the studies and more on its present day challenges.
- Shrink group size, keep presentations shorter. Loved the work, very informative. Recommend you do it again.
- The experience with jigsaw was interesting. I liked working with different people. The only thing I did not like was it was not mentioned in the syllabus. I did enjoy and got a lot of information out of it though.

Instructor's self-reflection and students' informal feedback lead to improvements to the fall 2009 jigsaw technique. In fall 2010 and 2011, the jigsaw group size was reduced to four persons per group from a six-person group. The jigsaw groups were formed in the second week of the semester; this gave students almost the entire semester to learn about their topic and work on their presentations. The jigsaw technique was included in the course syllabus and made part of students' assessment. To assess students' presentations, a rubric was designed which clearly stated instructor's expectations and it also served as a guideline for preparation. Participation and presentation on a jigsaw topic was worth 50 points and the student's score was added to his or her final grade. The class size increased from 24 students in fall 2009 to 34–38 students in fall 2010 and 2011. This lead to the formation of more jigsaw groups in fall 2010 and 2011 than in the 2009 class, and so more current topics were introduced, making the scope of learning wider and richer. In order to test the depth of understanding, each student in a group was asked to generate three multiple-choice questions from his or her subtopic. To make learning more interactive, the groups seek answers to these questions from their classmates. The presentations were made short by setting a 15-minute time limit for each presentation. To make sure that the content of presentations was informative and yet appropriate to the level of the class, the groups and the instructor met with each other multiple times. The above changes to the jigsaw technique were implemented in fall 2010 and 2011. The following are student evaluation comments of fall 2010 class:

- The jigsaw presentations at the end of the course helped me think more critically about the subject.

- Enjoyed this class! I also enjoyed participating in the jigsaw presentations. I gained a better understanding of my topic as well as others that were presented. Plus, it gave me practice for presentations in the future.
- I enjoyed the presentations at the end of the class as I feel it helped me to better understand the topics that were assigned.
- The group presentations were very helpful in understanding the course material.

The class of fall 2011 gave its informal reflective feedback on jigsaw experiences. All the comments were very positive. The following comments are representative:

- I think the jigsaw presentations were a great learning experience. The topics that were covered are ones we all heard about but with these presentations we were able to take a deeper look. These topics are essential for everyone to know about, but especially for graduating biology majors. We are the ones who need to educate the average person about all these topics relevant to our lives. Also, most people are not comfortable presenting in front of a room full of people, having your group members on your side gives you a sense of confidence.
- I really enjoyed the jigsaw presentations. It was a fantastic opportunity to practice to public speaking. It was also a great way to learn searching scientific literature related to the topic. I found that short presentations are easy to follow. The presentations helped alleviate fears that are caused by public speaking. Everybody in the class was very attentive and the classroom had a positive atmosphere overall.
- The jigsaw presentations gave me an opportunity to learn about a variety of subjects and also the opportunity to meet new people, I probably would not have had a chance to meet. The jigsaw presentations also relived the stress of presentation because it was a group work instead of individual presentation. I enjoyed listening to other students as well.
- I thought the jigsaw presentation was a good way to learn and integrate a wide variety of cell and molecular biology concepts. Beyond simply learning biological process, the presentations provided information in practical and relatable ways. I think everyone did a great job researching and presenting in large part due to low pressure nature of the assignment. The only suggestion I would make would be to extend the presentations by one more class period. Overall, I felt the assignment was enjoyable, interesting and effective.
- The jigsaw presentations were fun to do. I learned a lot about topics that are a part of my everyday life. The only thing that I had a problem with was finding time to meet with my group. Our schedules clashed with each other, work, classes etc. But other than that, I had a blast!
- I think jigsaw projects are good practice for grad-school or any upper levels where you need to convey information clearly and concisely.
- I really enjoyed the jigsaw presentations. Not only did it educate me on an important biological/ scientific topics, but, it gave me a confidence to speak in front of a large group.

It also gave me more practice using PowerPoint to give a presentation. I learned to speak from knowledge gained instead of reading a slide. I wish more classes incorporated jigsaw presentations.

- The jigsaw presentation project was fun, innovative method of learning that is typically absent in upper-level biology courses. I enjoyed learning about a multitude of biology topics. Random assignment of topics was a functional and fair way of delegating and segregating work.
- I think it is very beneficial to have each group cover a specific topic. This allowed me to gain more knowledge on the important topics. Because the group members were drawn at random, I worked with three other students I had not really talked to before. I liked this aspect because I was able to get to know other people. There were a few difficulties with communication, but this just made the final results more rewarding.

A second criterion used to assess the successful implementation of the technique is the comparison of the final grades. The final grades of fall 2009 class (control group) and fall 2010 and 2011 classes (experimental groups) were compared (Figure 8.1, 9. 118). However, grades per se do not measure the success of the technique, as various other factors could be responsible for better grades, but grades together with student evaluations and informal reflective feedback are good indicators of the success.

The number of students getting higher grades (A and B) increased from 2009 to 2010 and 2011 (Figure 8.1). Higher grades in fall 2010 and 2011 indicate that the students grasped the course material better. This could be, in part, due to implementation of improved jigsaw technique. My senior colleague, who was invited to the fall 2011 class for peer evaluation as well as to critique and evaluate jigsaw presentations, had overall very positive impressions about the implementation of the technique and my role as a jigsaw teacher. She remarked that I was an excellent role model as a presenter to my students during my 15-minute short presentation on "causes of cancer," before the students' presentations began. She also noticed that some student presenters were reading from PowerPoint slides instead of "talking" to the audience.

Next Steps

My colleague suggested that to improve presentation skills of students, there should be video demonstrations of good and bad teaching presentations and then students should be asked to critic these presentations. This activity will be done in the beginning of the semester so that students take mental note of good and bad presentation skills and have enough time to practice and thereby develop good presentation skills. To allow more time for discussion after the presentation, I will extend the question-and-answer session by five minutes. This would require extending the presentations from three class periods to four class periods.

Conclusions

The implementation of the jigsaw technique is motivating students to learn information about rich and seemingly complicated topics in cell and molecular biology. This cooperative technique creates a nonthreatening and noncompetitive active learning environment (Sandberg 1995) by requiring students to research, read, question, write, discuss, and present to their peers. This learning method meets most of the teaching, professional development, assessment, and inquiry and content standards set forth by the National Science Education Standards (NSES). Implementing this cooperative learning technique, the teaching standards emphasized by NSES were met as I shared the responsibility of learning with my students; guided each one of them through the active inquiry process; provided opportunities to share, discuss, and debate scientific knowledge within the groups as well as with the entire class. As a result, a community of learners with shared responsibilities and respect for each other was created. The professional development standards as envisioned by NSES have been also met because I have improved the jigsaw technique by making several changes to the technique implemented in fall 2009. These improvements are based on my reflections, students' comments and feedbacks. To demonstrate how theory of inquiry teaching is translated into practice, I have shared with my colleagues my knowledge and practical experiences gained while implementing this technique by giving presentations at regional and national-level professional development conferences.

The students were assessed according to the assessment standards recommended by NSES. A rubric was designed by me to assess students' performance. To make the assessment expectations clear, the rubric was discussed and given to the students at the beginning of the semester. Students were expected to create annotated bibliography to assess their ability to investigate scientific literature. They were asked to give oral presentations, generate multiple-choice questions and answer their classmates' questions at the end of the presentation. These activities tested students' understanding and their ability to reason and communicate scientific information.

The jigsaw learning method met content and inquiry standards of NSES. The topics covered using this technique may be controversial, multidimensional, global, social, or on the forefront of research and so students could relate to these topics and felt motivated to explore. They analyzed, understood, and organized the information in the form of PowerPoint presentation and communicated to their group members and classmates by giving oral presentations.

Many of my students have developed interest in the discipline as a result of jigsaw, and some of them are currently pursuing graduate studies in the discipline. I will continue to use this cooperative learning technique in this upper-level college course to make learning permanent.

References

Alico, R. 1997. Enhancing the learning in microbiology through cooperative learning. Paper presented at the annual meeting of the Mid-Atlantic Association of Microbiologists, Montgomery, MD.

Aronson, E., and S. Patnoe. 1997. *The jigsaw classroom: Building cooperation in the classroom.* New York: Addison Wesley Longman.

Baviskar, S. N., R. T. Hartle, and T. Whitney. 2009. Essential criteria to characterize constructivist teaching: Derived from a review of the literature and applied to five constructivist-teaching method articles. *International Journal of Science Education* 31 (4): 541–550.

Cooper, J. 1990. Cooperative learning and college teaching: Tips from the trenches. *Teaching Professor* 4 (5): 1–2.

Heady J. 1993. Teaching embryology without lectures and without traditional laboratories: An adventure in innovation. *Journal of College Science Teaching* 23 (2): 87–91.

Johnson, D., R. Johnson, and K. Smith. 1991. *Active learning: Cooperation in the college classroom.* Edina, MN: Interaction Book Company.

Kagan, S. 1994. *Cooperative learning.* San Clemente, CA: Kagan Publishing.

Kohl, H. 1982. *Basic skills.* New York: Little, Brown.

Lord, T. R. 2001. 101 Reasons for using cooperative learning in biology teaching. *The American Biology Teacher* 63 (1): 30–38.

Maloof, J. 2004. *Using the jigsaw method of cooperative learning to teach from primary sources. www.doit.gmu.edu/inventio/issue*

McKeachie, W. J., P. R. Pintrich, Y.-G. Yin, and D. A. F. Smith. 1986. Teaching and learning in the college classroom: A review of the research literature. Paper presented at the National Center for Research to Improve Postsecondary Teaching and Learning, Ann Arbor, Michigan.

Meyers, C., and T. B. Jones. 1993. *Promoting active learning: Strategies for the college classroom.* San Francisco: Jossey-Bass.

Sandberg, K. E. 1995. Affective and cognitive features of collaborative learning. In *Review of research and developmental education,* ed. G. Kierstons, 6. Boone, NC: Appalachian State University.

Slavin, R. E. 1990. *Cooperative learning-theory, research, and practice.* Englewood Cliffs, NJ: Prentice Hall.

Stahle, R. J., and R. L. VanSickle. 1992. Cooperative learning as effective social study within the social studies classroom. Paper presented at the National Council for Social Studies, Washington, DC.

Table 8.1. Rubric for Assessment of Jigsaw Presentation

Criteria/ Component	10 points	5 points	2 points	Total	Comments
Information	Accurate and complete information presented in a concise, logical sequence, like a story. Presenter interacted and made eye contact with the audience.	Information is incomplete. Lengthy text (not concise). Presenter mostly read the information from the slides, with little explanation. Little interaction or eye contact with the audience.	Major concepts or information is missing; logical sequence is not evident. Presenter read the information from the slides. No Interaction.	—	
Background and text, spelling, punctuation, and grammar	Background and text complement each other; easy to read, consistent throughout the presentation (title of slide, text, background color, placement & size of graphic, fonts—color, size, type for text and headings) correct grammar usage is evident.	Background is not consistent throughout the presentation; text size and color change with each slide; a few grammar and spelling mistakes are present.	Text cannot be read on selected background; text size and color make it difficult to focus on information. Spelling mistakes occur throughout the entire presentation; standard grammar usage is not evident.	—	

Continued on next page

Continued from previous page

Criteria/ component	10 points	5 points	2 points	Total	Comments
Graphics and transitions	Graphics (pictures, graphs, charts, animations, movies etc.) are appropriate and relate to content; Enhance understanding of concepts and their relationships. Transitions are consistent throughout the presentation. Sources of graphics cited at the bottom of the slide.	Few graphics are used throughout the presentation, graphics do not complement well to the topic presented; transitions are not consistent or effective.	Little or no attempt was made to use graphics or images too large/too small, distracting, create a busy feeling.	___	
Annotated bibliography	A completed and accurate bibliography is included. Shows amount of work and depth of understanding	Very few citations included; annotations very brief, display superficial understanding and a mediocre preparation.	Bibliography is not included.	___	
Multiple-choice questions	Questions test knowledge and understanding. Used simple, precise, and unambiguous wording.	Questions do not test both knowledge and understanding.	Questions test only facts (easy); ambiguous.	___	

Figure 8.1. Final Grades of Fall 2009 Class Were Compared With the Final Grades of Fall 2010 and Fall 2011 Classes After Implementation of Jigsaw Technique in Cell and Molecular Biology Course.

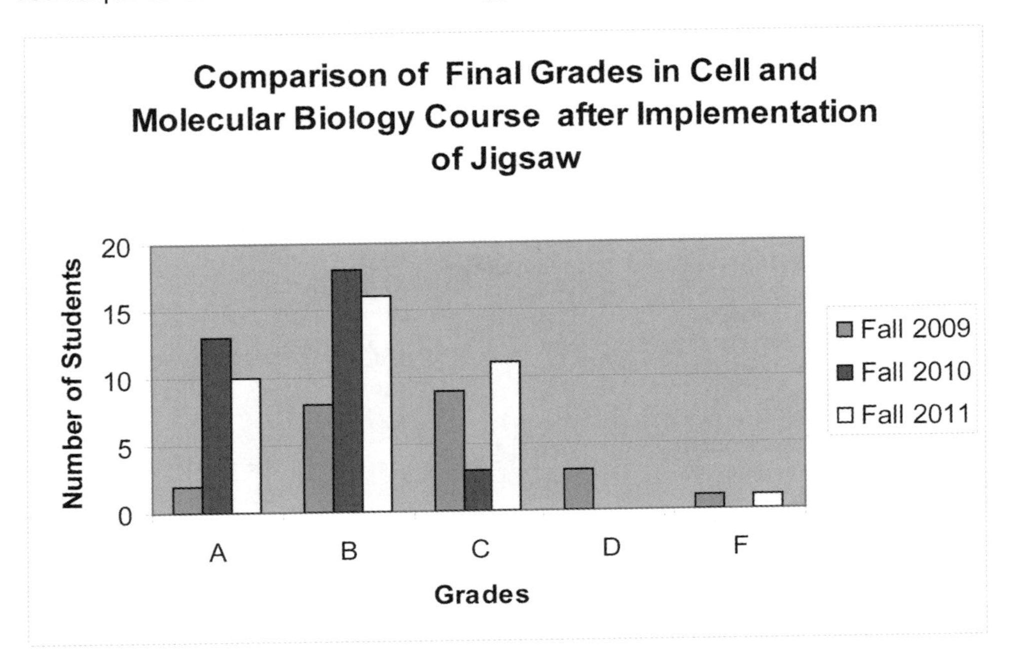

Clickers in the Geoscience Classroom: Pedagogical and Practical Considerations

David N. Steer
University of Akron

Setting

Rapid advances in computing hardware and software have resulted in widespread adoption of automated classroom feedback systems, also known as clicker systems. Faculty members at The University of Akron have been using these systems for rapid formative feedback in classes of all sizes for over a decade. This chapter reports on the use of such personal response systems in general education Earth science classes for nonmajors. It reviews the impact on student learning, lessons learned, and insights for the future based on data collected over the past five years.

Overview: Clickers—What Are They and Why Use Them?

There are several versions of modern student response systems, but they all share some common elements (Barber and Njus 2007). Students purchase a remote control through a commercial vendor or are provided one as part of a course. Generally these remotes are registered by the student through that vendor's website. Once registered, students use the remotes to input answers to questions. Those data are then available in near-real time for display as a histogram on the classroom screen. Data are archived and available for instructor review or analysis. The instructor part of the system includes a receiver and software provided by the same vendor that supplies the remote control units. Some systems require multiple receivers with line-of-sight to infrared transmitters. Those systems have generally been replaced with radio-frequency transmitters without such constraints. The software is usually designed to incorporate PowerPoint slides, into which instructors can insert multiple-choice questions as a routine part of the lecture. Most systems are limited to simple letter or numerical input.

While some faculty members may consider the use of student personal response systems as innovative, the opposite is true. Early versions of this technology were introduced in the 1960s when mainframe computing first became available on campuses across the country (Judson and Sawada 2002). These early systems looked nothing like today's modern analogs, but they served essentially the same purpose. That technology was hardwired into the desks with responses transmitted only to the instructor, who then informed students about class response patterns (Judson and Sawada 2002). Early attempts at using in-class technique were quickly abandoned due to the high cost of implementation and limited effectiveness at improving student learning (Bessler 1971; Brown 1972; Bapst 1971; Casanova 1971). So what changed that makes such systems better today than in the past?

In their review of previous attempts to use this technology in classrooms, Judson and Sawada (2002) noted that this pedagogical approach failed because it did not become a vehicle for curricular reform. Rather, faculty members used the systems to simply ask low-cognitive-level questions (Anderson and Krathwohl 2001) that students could answer (or not) based on their attention level. Little attempt was made to use the technology to engage students in a more collaborative learning environment that required higher-level thought. Several things changed in the 1990s that caused automated in-class assessment to gain favor in classrooms across the country. First, the cost of the technology dropped dramatically from that of the early hardwired prototypes that required servers (Judson and Sawada 2002). Second, the hardware and software became much easier for students and faculty to use. At the same time, the software included tools that facilitated interactions and data analyses (Roschelle, Penuel, and Abrahamson 2004). Third, curricular reform efforts in the sciences gained national attention as landmark studies endorsed student-centered pedagogical approaches that engaged students with one another and their instructors (American Geophysical Union 1994; NSF 1996; NRC 1997 and 2000). Teaching reform efforts coupled with science education research greatly expanded in subsequent years. This resulted in guiding principles that facilitate learning when adopting this in-class, technology-enhanced pedagogical approach (Mazur 1997; McConnell, Steer, and Owens 2003; Greer and Heaney 2004; Caldwell 2007; Beatty and Gerace 2009).

Instructional Program: Clickers—How Should They Be Used?

The nature and level of student-student interaction facilitated by this technology plays a critical role in the overall effectiveness of this approach. One of the simplest techniques to increase student interaction is the peer-instruction method introduced by Mazur (1997) in his physics classes at Harvard. Mazur posed a multiple-choice conceptual question (called a ConcepTest) for students to answer (first using scan cards, later using clickers). Students discussed their answers with one another and had an opportunity to answer the question again after that conversation. With the advent of modern systems that display a student-response histogram in near-real time, that peer discussion is now often guided by that graph of student responses. Modifications of Mazur's procedure include variations such as asking different, but conceptually equivalent follow-up questions or peer-interaction before answering the first question (Nichol and Boyle 2003). In any case, the technology becomes a facilitator of interactions as students discuss questions that require deep thought to improve their learning (Mayer et al. 2009).

There are several guiding principles related to ConcepTest question design that can influence learning outcomes. First, the level of the learning objective probed must be assessable using a multiple-choice question that generates discussion. Remembering-level questions (Anderson and Krathwohl 2001) that simply assess rote memorization are ill-suited for this approach. Such questions do not encourage the student to think and respond, rather they simply require the student access their short- or long-term memory (NRC 2000). Such questions short-circuit peer-discussion because students have nothing to discuss except the meaning of a particular word or phrase. Recall items are not formative other than to inform a student that he or she does not know a necessary definition. Better questions build requisite vocabulary into a question where discussion is beneficial. Questions that touch understanding through analyzing cognitive levels (Anderson and Krathwohl 2001) are much more powerful. More advanced questions provide an opportunity for students to self-assess as they grapple with concepts that have been recast in new ways that require high-level thinking (Aiken 1982). These response systems provide a low-stakes method where students practice thinking about how much they know and why they know it when concepts are rearticulated in new ways. These opportunities for self-reflection play a critical role in student metacognition (NRC 2000). With clickers, students have an opportunity to affirm their new level of understanding when instructors present follow-up questions that build deeper conceptual connections.

Question format also contributes to the overall effectiveness of this approach. To the degree possible, each question should focus on a single concept deemed essential to student success in the course. The question should use simple language, uncomplicated graphics (Gray et al. 2011), avoid direct calculations, and include known alternative conceptions (McConnell, Steer, and Owens 2003). This in not to say the question cannot involve multiple concepts; rather that the best questions for discussion relate to a single idea because that will focus student discussions (Caldwell 2007). For example, a geoscience student might be required to understand that the process of plate tectonics leads to distinct patterns in seafloor age data. The processes that constitute the theory of plate tectonics are directly linked to the characteristic patterns in the data. Conceptual understanding of this connection can be probed by asking the student to select a graph that illustrates the relationship between seafloor age and distance from a mid-ocean ridge where plates form (Figure 9.1a, p. 122). There are two main ways students might approach answering this question. Students could simply recall their mental image of the age of the seafloor map or relate it to the processes forming new seafloor. If this question had text answers (e.g., seafloor gets older moving away from the ridge), it would be a low-level memory question and probably would not merit discussion. In this case students must analyze graphs they have not seen before, thus making the question more effective.

Figure 9.1a. Analysis-Level ConcepTest Question

This question requires that students be able to interpret a graph that recasts data typically viewed in map view as a color image. Correct answer is A. Student responses indicated students understood the relationships quite well after completing their homework with 80% selecting A, 2% B, and 18% C.

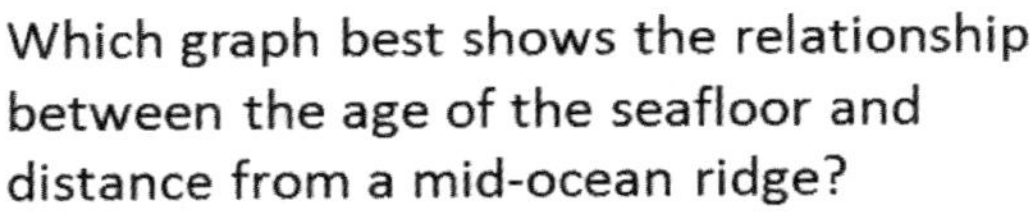

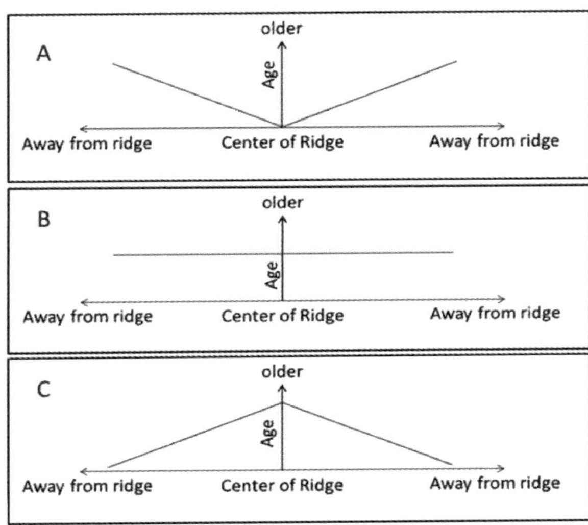

Such a question does not address the link between the graphs and the processes that cause the patterns. These connections can be made through follow-up questions where students are asked to interpret the shape of the lines or curves placed on the graph and link that to process (Figure 9.1b). In this case, the change in the slope of the curves suggests plate motion changed over time. This is a more difficult conceptual question that requires a higher level of understanding. In this case, students had difficulty interpreting the meaning of the curve in the graphs. Even after discussion, about half of the students thought the graph implied increased plate motion; about half thought the curve meant decreased velocity. The instructor had to intervene and explain why the graph illustrated slower plate motion in the past. This example illustrated that instructor-student interaction may be required to deconstruct the question if peer instruction does not sufficiently improve the response pattern.

Figure 9.1b. Analysis-Level ConcepTest Question

This question requires that students be able to interpret a graph and link that interpretation to the process of plate tectonics. Correct answer is A. Student responses to this question varied, with 45% selecting A, 3% B, 47% C and 5% D. In this case, postdiscussion answers migrated only slightly with no students selecting B or D and 55% moving toward the correct answer.

What does the shape of this curve tell you about the plate forming process at this location?

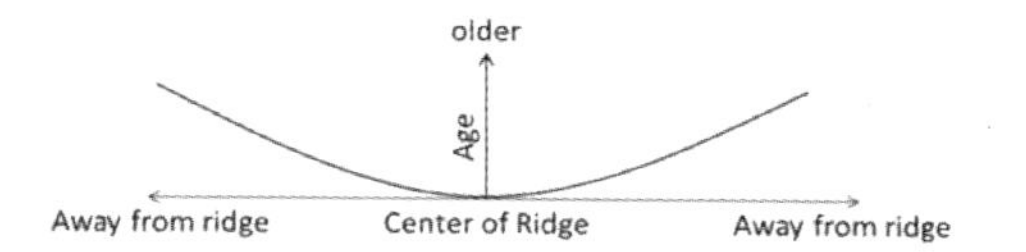

A. Oceanic plates formed more slowly in the past at this location.
B. Continental plates formed more slowly in the past at this location.
C. Oceanic plates formed more quickly in the past at this location.
D. Continental plates formed more quickly in the past at this location.

Evidence for Success—Assessing the Impact on Students

Students' responses to conceptual questions answered using personal response systems were collected in general education Earth Science classes for nonmajors over a five-year period. The setting was in an urban environment primarily serving the needs of students in the adjoining six counties. Classes typically consisted of 160 students in a lecture-style auditorium with fixed seating facing a large projection screen. Student demographics of the classes each year generally mirrored that of incoming students to the university with a nearly 1:1 female-male ratio. Of this, 70% were Caucasian, 19% African American, 3% Hispanic and 8% listed as other categories. Because this was an entry-level general education course with no prerequisites, 65% of the students were freshman and 25% were sophomores, with the remaining 10% categorized as other (e.g., secondary school students, upper-classmen/women).

Prior to class, students completed a mediated instruction of text homework assignment (Neal and Langer 1992) that was turned in for a grade upon entering the classroom. Completion of these assignments was a critical component of the instruction because in-class activities built upon concepts covered in the text and students frequently do not know how to effectively use their textbooks (Neal and Langer 1992). The homework consisted of two or three pages of short-answer questions that could generally be answered after reading the material and responding to the prompt with one or two sentences. Student responses to the questions on these assignments served as notes that identified key concepts, vocabulary, and ideas in the reading. For instance, "List and define the five principal components of the Earth System" was a low-level, content-type prompt. A question that required more sophisticated analysis of the material was phrased like, "Read through the section entitled 'The Anthropocene: A new time on Earth?' Why is the section given that title?" When properly completed, the homework served two purposes. When preparing for exams, students did not have to refer to the text because they had a record of the main ideas covered in the reading. These assignments also prepared students for class, where basic concepts in the reading were expanded or applied during activities rather than being repeated or defined.

Students entered the classroom and were given a variety of materials. A teaching assistant handed out copies of selected PowerPoint slides, occasionally other curricular materials needed for the lesson (e.g., models, rock samples, globes) and a one- or two-page lecture-tutorial worksheet (Kortz and Smay 2010). Students were provided essential PowerPoint slides so they did not have to copy slide text. This allowed them to focus on the instructor discussion. Students were encouraged to take notes about the slides to better clarify the concepts or emphasize the most important concepts discussed. Students sat in randomly organized, permanent, four-person learning teams with assigned seating for the entire semester. They were required to have a registered, functional remote by the fifth lesson. In class, they worked together as they responded to conceptual questions and completed their lecture tutorial, which included a variety of formative assessments (e.g., graphical analysis, Venn diagram, concept map). The instructor would cover a learning objective and then pose a ConcepTest question or have students complete a related activity (Gray et al. 2010). Lecture by the instructor was secondary to these activities that were sequenced, varied, and spaced throughout the entire lesson.

This study focused on longitudinal data collected over this period from 2,359 students taking the same class taught by the same instructor every semester. Student demographic data, attendance, weekly quiz scores, attitudinal surveys, and course grades were compared to correct answer score rates on ConcepTest questions asked in class throughout the semester. Partial data sets from students who did not complete the course were excluded. The Geoscience Concept Inventory (Libarkin and Anderson 2006) was used as an external pre- and postcourse assessment of geoscience conceptual understanding. Average student scores by lesson and coursewide were compared qualitatively or using single factor ANOVA, correlation analysis or *t*-tests where appropriate. Trained graduate student assistants observed student group interactions during one semester. All students reported herein agreed to this research by signing informed consent that allowed collection and analyses of the data.

As expected, clicker data confirmed that final course grades were strongly correlated with attendance as measured by the number of clicker responses received. The main advantage of using electronic means to record attendance was speed and the ease with which that data could be processed. Once students turned on the remote control and answered a question, the system logged their participation and attendance. Students were allowed to write down their answers if they forgot their clicker or it malfunctioned. Prior to the introduction of clickers, taking attendance was difficult and time-consuming. A teaching assistant manually recorded attendance based on whether a seat was occupied or not. That TA was then unavailable to assist in learning activities. Students attended those instructor-focused lecture sections at rates that varied from 40–75% over the course of a semester; typical of many passive science lecture courses (Wren 2011). Once clickers were introduced, median attendance improved to an average of 83% (median was used here because the distribution was forward skewed).

Since the importance of attendance in a college course is well established (Massingham and Herrington 2006; Crede, Roch, and Kieszczynka 2010), a grade was associated with student participation in this course. If students missed class, they lost points (unless the absence was excused). That 10–20% participation grade was based partly on simple attendance and partly on actually answering questions throughout the course. Others variations to encourage attendance included using these systems to award extra credit or only when correct answers were submitted. While some cheating was observed, our experience suggested this rarely occurred because physical attendance was randomly verified by collecting class exercises. The instructor and teaching assistant could also observe if a student used more than one device as they were circulating through the room. Our data suggested that positive link between final grades and attendance was established almost immediately when using clickers (R^2 = 0.5 on the first lesson used, 0.4 after two lessons). That trend continued through the semester (R^2 = 0.3). Such early information that was easily collected using a personal response system provided the instructor an early warning flag that was used to try to intervene with those students. Unfortunately, many students who were contacted regarding their poor attendance did not change their behavior.

The precourse geoscience concept inventory (GCI) was a useful predictor of how well students would answer in-class conceptual questions. The GCI is a 12-question valid and reliable multiple-choice format instrument for measuring student conceptual understanding for a variety of geologic concepts (Libarkin and Anderson 2006). When students who had a GCI score were binned into thirds by score, the lower one-third of students (n=1197: GCI score of 1–4) correctly answered an average of 49% +/– 12% on the in-class ConceptTests. Students in the middle (n=1544: 5–8 GCI) and upper thirds (n=163: 9–12 GCI) scored 58% +/– 14% and 59% +/– 19% respectively. That result mirrored previous work describing the impact of prior knowledge on student performance in a variety of disciplines (Symons and Pressley 1993; NRC 2000; Hewson and Hewson 2006; Hailikari, Katajavuori, and Lindblom-Ylanne 2008). There were students in the low GCI pretest category that scored very well on the in-class ConcepTests and some in the high GCI group who scored low on those questions (Figure 9.2, p. 126). That was explained by analyzing attendance of those students. Students who performed low on the GCI, but attended regularly did better than higher GCI-scoring students who attended less regularly (~10% difference in attendance, $p < 0.001$).

Figure 9.2. Comparison of Precourse GCI to Average Correct Scores on ConcepTests

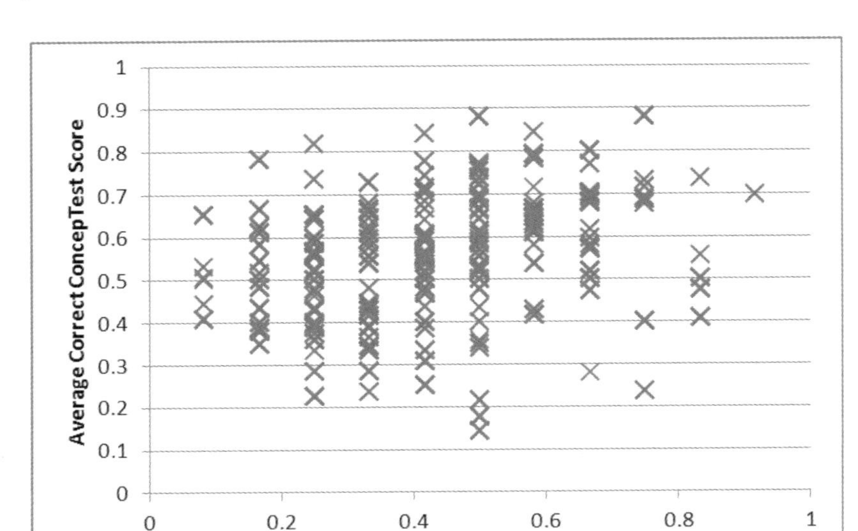

Female student correct answer rates on in-class questions were about 5% higher than male students' scores (56 +/– 15% versus 50 +/– 13% respectively; $p < 0.001$). This equated to female students answering one more question correctly than their male counterparts every four to five lessons. That difference, while statistically significant, would not be discernible to the students or instructor during the normal course of instruction. However, the distribution of female correct answer rates was more forwardly skewed to higher rates than that of males (Table 9.1). This was consistent with research showing males' multiple-choice scores displaying a larger spread than those of females (Cole 1997). The finding of nearly equal performance levels found here supports the contention that the clicker pedagogical approach is equally inclusive for males and females (Steer et al. 2009).

Table 9.1. Percent Correct Answers

Rate	Female	Male
<25%	4%	2%
25-50%	24%	40%
50-75%	62%	56%
75-100%	9%	2%
	n=2054	n=1334

Caucasian students were significantly more likely to correctly answer ConcepTest questions than their minority counterparts (56 +/– 14% versus 44 +/– 12% respectively; $p < 0.001$). This finding was partially explained by differences in attendance and pre-course GCI scores. Minority students in these classes attended less frequently than Caucasian students (74% versus 80% attendance). Their GCI scores were also lower by an average of 1 answer out of 12 on the GCI. The combination of lower attendance and less precourse conceptual understanding likely explained the differences in overall classroom performance on these questions. Again, while this difference in correct answer rates was noticed statistically, it was unlikely these differences would be noticed during instruction. The difference in correct answer rates equated to minority students answering one more question incorrectly than their Caucasian counterparts every two or three lessons.

There were several notable findings from an attitudinal survey (Table 9.2, p. 128) administered one semester ($n = 88$). That survey confirmed other studies that showed students valued the direct feedback provided by clickers (Abrahamson 1999; Beekes 2006). Over 70% of students responded that they agreed or strongly agreed with statements such as "I think the clicker system helps me learn," "do better in the course," and "I think carefully about my response before I answer." Around 80% answered in the same fashion when presented with questions asking if they were interested in how their answer compared to the rest of the class and if they defended their answers. Inverted forms of the same questions confirmed these findings (Table 9.2, p. 128, items 2–9). Unfortunately, students were neutral to the statement "I make a note to help me remember the concept when I miss a question." This pointed out one major weakness in using ConcepTest questions in these classes. This pedagogy was built on the paradigm that frequent, formative feedback is essential for student metacognition and learning (Daws and Singh 1996; Black and Wiliam 1998). When students answer the questions, they are provided opportunities to assess their own learning. If students are not using the information to remind themselves that they had difficulty in class with a concept, it is unlikely they will remember to refresh themselves more thoroughly when preparing for an exam. A qualitative comparison of student in-class responses to similar, but different exam questions covering the most difficult topics appeared to confirm this disconnect.

Analyses of question-response patterns highlighted student progress and topical areas more conceptually difficult for students to grasp. Mazur (1997) suggested correct answers on the first attempt should fall in the 30–70% correct range for effective peer instruction to occur. Meaningful discussion leading to improved learning was not likely to occur if too few students answered correctly on the first attempt. If the large majority of students answered correctly on the first attempt, there was little value in spending additional time on that concept. Our analyses of pre-and-post-peer instruction data showed average normalized gains of 40% across a large spectrum of initial response rates (Figure 9.3, p. 129). These data suggested students benefited almost regardless of the initial score on a question.

Table 9.2. Survey Results (n = 88)

Survey Question Order	Question	Strongly Disagree	Disagree	Neither Agree nor Disagree	Agree	Strongly Agree
1	I think the clicker system helps me learn.	7%	6%	12%	53%	22%
3	I am interested in seeing how my answer compares to the entire class.	6%	4%	8%	54%	28%
6	Clicker questions help me do better in class.	7%	7%	16%	49%	22%
7	I think carefully about my response before I answer.	1%	6%	9%	58%	26%
2	Viewing how the whole class responded is not helpful in my learning.	11%	46%	29%	4%	9%
4	I don't worry too much about it when I miss a question the second time.	17%	58%	15%	8%	2%
5	Given the opportunity to answer a second time, I usually just ask a neighbor.	16%	47%	15%	19%	3%
9	I often just guess which answer is correct.	17%	45%	20%	15%	2%
8	I make a note to help me remember the concept when I miss a question.	7%	33%	20%	33%	8%
10	I defend my answer with my group before answering a second time.	1%	6%	15%	64%	13%

Figure 9.3. Pre- Versus Postdiscussion Answer Response

Comparison of average pre–peer discussion question to postdiscussion questions for questions asked every other lesson over a semester. Points above the black line showed an improvement in student correct answer rate after peer instruction.

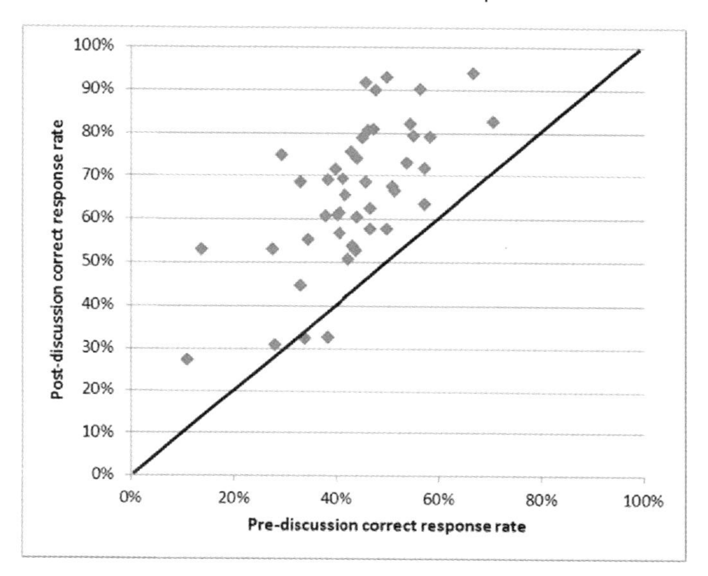

Pre- versus postdiscussion correct answer rates organized by lesson displayed other trends (Figure 9.4, p. 130). While most initial responses fell in the 35–55% range, postdiscussion gains varied by topic. Low post–peer instruction gains were associated with concepts already known to be conceptually difficult for students. Those topics included plate tectonics (Barrow and Haskins 1996; Libarkin and Anderson 2005; Sibley 2005; Steer et al. 2005), geologic time and numerical change estimates (Libarkin, Kurdziel, and Anderson 2007), density-related topics (Yeend, Loverude, and Gonzales 2001) and those involving math (Ballard and Johnson 2004). Questions probing other concepts such as earthquakes, volcanoes, weather, and climate tended to show larger gains.

Figure 9.4. Pre and Post–Peer Instruction Scores by Topical Area

These questions showed stable response patterns over the semesters. Small gains were associated with more difficult concepts though almost all questions showed some improvement after peer instruction.

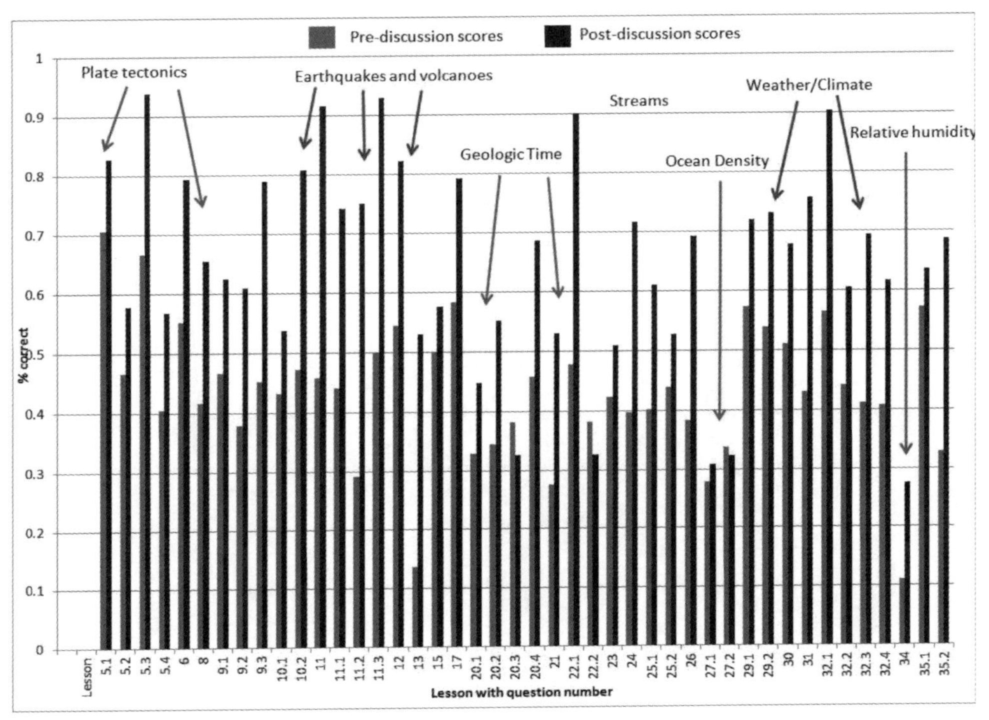

Qualitative data collected by student research assistants observing group dynamics highlighted some challenges related to using this pedagogical approach. Trained graduate assistants recorded observations of group interactions on multiple lessons covering a variety of topics. Three general findings related to clicker pedagogy could be distilled from reviewing these written observations. When group dynamics were good, students learned from one another though sometimes this led to misinformation. In one instance, students were presented with a diagram and asked the question "Select the letter where epicenters with the deepest foci are located. Choose A, B, C or D" (Figure 9.5). Results of that initial polling were heavily, but equally weighted toward two of the four possible choices (A and B).

Figure 9.5. Schematic Map Used to Ask Various ConceptTest Questions Related to Plate Tectonics

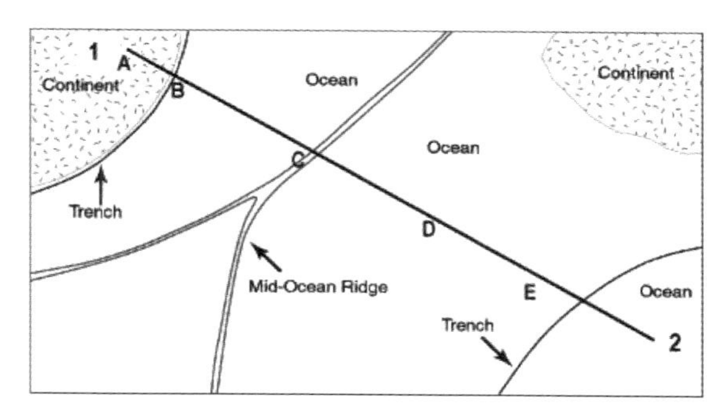

Research Assistant Jill recorded the following four-person group conversation related to this question:

Student 1: "I think B, that red line in the trench, *sic* but earthquakes aren't in the ocean."

Student 3: "OK, if plate goes under, earthquakes are in that plate"

Student 1: "Why not B, why aren't they there?"

Student 2: "They just don't. The one under is the one melting, changing, that's where the earthquakes are."

Student 1: "Makes sense."

Student 2: "A and B looks like the same thing. I ... like bigger are deeper, right? I don't want to tell you the wrong thing."

Direct Instructor input to class: "Big earthquakes are shallow and deep, here we ask for depth. Think about the depth issue."

Student 3: "It the location of the foci"

Student 2: "Oh, I get it, how deep they are ... it's A"

In this example, the students collectively came to the correct conclusion in this less-than-one-minute conversation. This conversation also illustrated how incorrect ideas can be brought up in the course of discussing a question. Student 1 thought the deepest earthquakes occurred at the oceanic trench and student 2 thought deeper earthquakes were bigger. Both ideas were incorrect. Fortunately, the instructor noticed that student discussions related to earthquake size were a point of confusion throughout the class and intervened. Since the question dealt directly with the depth distribution of earthquakes, the group was able to reject option B and move toward the correct answer. However, the size versus depth misconception had to be addressed separately.

Faculty monitoring such conversations must keep current on misconception literature and be prepared to address such issues where they might arise.

In other cases, group dynamics were not always beneficial. Bill, the graduate assistant observer, noted this interchange between students discussing a question related to planets:

Student 1: "Do you know the answer?"

Student 2: "I ... don't really know." (Observer note: Not much group talk, student 3 sits and watches, student 4 works alone.)

Student 2: "Any ideas?" (Observer note: Student 4 shows paper.)

Student 2: Nods head. "Good guess anyway."

All students re-answer question.

In this example students were apparently not engaged by the subject, which resulted in less useful conversations that likely limited learning. Students were only concerned about the answer, not the reasons or concepts behind that answer.

The Future of Personal Response Systems

This approach and the devices that enable it are evolving. As technology changes, there may be no need for the types of devices we use today or the input systems may provide richer response options. However as the systems change, there will be a continuing need to document learning gains. Findings from collaborators in this research project suggested faculty sometimes become disenchanted with the systems if they observe no significant change in overall student outcomes (Semken 2010). Indeed, some faculty members in this department stopped using clickers after several semesters when, in their view, the costs of the clickers did not outweigh the benefits. Likewise, institutional support is important for this approach to gain wide acceptance at a collegewide level (Armour 2010). Innovative methods to use the systems will determine future viability of this approach. For instance, student performance greatly improved when in-class questions were posted online and students were allowed to re-answer them until they scored 100% (personal communication, Anne Holmes 2012). Knott (2010) also discovered that students who used clickers showed greater improvement in their post-course GCI scores than students who did not use the technology. The positive indications of using this approach were also documented by Kortz (2010) who found correlations between in-class performance and exam scores for community colleges.

While this pedagogical approach is now relatively simple for faculty to implement and is accepted by students, there are inherent limitations to asking only multiple-choice questions. Such questions generally reach only low-to-moderate cognitive levels and the repeated use of one assessment mode in class may lead to student disinterest. There are now commercial systems that use phones or laptops for the data entry device. Future advances in technologies may even make this approach obsolete. Some systems may eventually allow students to draw sketches or answer using text that can be quickly scanned for analysis. Regardless of the system, the basic design must include an approach that better facilitates metacognition to have the most positive impact on student learning. This pedagogical technique, like any other, works best when combined with a variety of learning activities designed to stimulate discussion and focus student attention on core ideas.

Other Reform Efforts

This pedagogical approach and technology can facilitate the developing national interest to document student learning gains in courses and programs across the sciences. Many science-related disciplines have or are developing literacy documents that focus on learning in higher education. For instance, in mathematics, core competencies are collectively called *quantitative literacy* (Madison and Steen 2008). Central ideas necessary for citizens to understand how chemistry affects humans are *chemical literacies* (Gewalt and Adams 2011). *Geosciences literacies* focus on content areas of climate, the atmosphere, the oceans, and the solid earth, with additional emphasis on grand challenges facing society (Zoback 2001). A common attribute found in these literacies is their emphasis on addressing how the subject area impacts the human condition and major interdisciplinary societal problems. While these data collection systems are not yet at a level where they can easily facilitate critical thinking, they are effective for measuring conceptual understanding of core concepts. ConcepTest questions can be developed that assess these core science concepts that exist or as they are elucidated. Such data could provide evidence of change in students' views of the world. Likewise, more and more departments are taking an interest in documenting student conceptual change at the program level. Data collected early in students' academic careers could easily be archived and then compared to later work throughout the program to document conceptual change using these systems.

Note: This research was supported through NSF grant DUE-0716397. Any opinions, findings, and conclusions or recommendations expressed in this material are those of the author(s) and do not necessarily reflect the views of the National Science Foundation.

References

Abrahamson, A. L. 1999. Teaching with a classroom communication system: What it involves and why it works. Mini-course presented at the VII Taller Internacional Nuevas Tendencias en la Ensenanza de la Fisica, Benemerita Universidad Autonoma de Puebla, Puebla, Mexico. *www.bedu.com/Publications/PueblaFinal2.pdf*

Aiken, L. R. 1982. Writing multiple-choice items to measure higher-order educational objectives. *Educational and Psychological Measurement* 42: 803–806.

American Geophysical Union. 1994. Report of the AGU Chapman Conference on scrutiny of undergraduate geoscience education.

Anderson, L. W., and D. R. Krathwohl, eds. 2001. *A taxonomy for learning, teaching, and assessing: A revision of Bloom's taxonomy of educational objectives.* Boston: Allyn and Bacon.

Armour, J. 2010. Choosing a classroom response system for a university-wide adoption: What's the big deal? *Geological Society of America Abstracts with Programs* 42 (5): 556.

Ballard, C., and V. Johnson. 2004. Basic math skills and performance in an introductory economics class. *The Journal of Economic Education* 35 (1): 3–23.

Bapst, J. J. 1971. The effect of systematic student response upon teaching behavior. PhD diss, University of Washington.

Barber, M., and D. Njus. 2007. Clicker evolution: Seeking intelligent design. *CBE Life Sciences Education* 6: 1–8.

Barrow, L. H., and S. S. Haskins. 1996. Earthquake knowledge and experiences of introductory geology students. *Journal of College Science Teaching* 26: 143–146.

Beatty, I., and W. Garace. 2009. Technology-enhanced formative assessment: A research-based pedagogy for teaching science with classroom response technology. *Journal of Science Education Technology* 18: 146–162.

Beekes, W. 2006. The "millionaire" method for encouraging participation. *Active Learning in Higher Education* 7 (1): 25–36.

Bessler, W. C. 1971. The effectiveness of an electronic student response system in teaching biology to the non-major utilizing nine group-paced, linear programs. PhD diss, Ball State University.

Black, P., and D. Wiliam. 1998. Inside the black box: Raising standards through classroom assessment. *Phi Delta Kappa* Oct: 1–13.

Brown, J. D. 1972. An evaluation of the Spitz student response system in teaching a course in logical and mathematical concepts. *Journal of Experimental Education* 40 (3): 12–20.

Caldwell, J. 2007. Clickers in the large classroom: Current research and best-practice tips. *CBE Life Sciences Education* 6 (1): 9–20.

Casanova, J. 1971. An instructional experiment in organic chemistry: The use of a student response system. *Journal of Chemical Education* 48 (7): 453–455.

Cole, N. 1997. The ETS gender study: How females and males perform in educational settings, ETS Technical Report. Available online at *ftp://etsis1.ets.org/pub/res/gender.pdf*

Crede, M., S. Roch, and U. Kieszczynka. 2010. Class attendance in college: A meta-analytic review of the relationship of class attendance with grades and student characteristics. *Review of Educational Research* 80 (2): 272–295.

Crouch, C. H., and E. Mazur. 2001. Peer instruction: Ten years' experience and results. *American Journal of Physics* 69 (9): 970–77.

Daws, N., and B. Singh. 1996. Formative assessment: To what extent is its potential to enhance pupils' science being realized? *School Science Review* 77: 99.

Gewalt, E., and B. Adams. 2011. A chemical information literacy program for first-year students. *Journal of Chemical Education* 88 (4): 402–407.

Gray, K., K. Owens, X. Liang, and D. Steer. 2011. Assessing multimedia influences on student responses using a personal response system. *Journal of Science Education Technology* (May): 1–11.

Gray, K., D. Steer, D. McConnell, and K. Owens. 2010. Using a student-manipulated model to enhance student learning in a large, lecture class: Addressing common misconceptions pertaining to the causes for the seasons. *Journal of College Science Teaching* 40: 86–95.

Greer, L., and P. J. Heaney. 2004. Real-time analysis of student comprehension: An assessment of electronic student response technology in an introductory earth science course. *Journal of Geoscience Education* 52 (4): 345–51.

Hailikari, T., N. Katajavuori, and S. Lindblom-Ylanne. 2008. The relevance of prior knowledge in learning and instructional design. *American Journal of Pharmaceutical Education* 72 (5): 113.

Hatch, J., M. Jensen, and R. Moore. 2005. Manna from heaven or "clickers" from hell: Experiences with an electronic response system. *Journal of College Science Teaching* 34 (7): 36–42.

Hewson, M., and P. Hewson. 2006. Effect of instruction using students' prior knowledge and conceptual change strategies on science learning. *Journal of Research in Science Teaching* 20 (8): 731–743.

Judson, E., and D. Sawada. 2002. Learning from past and present: Electronic response systems in college lecture halls. *Journal of Computers in Mathematics and Science Teaching* 21 (2): 167–81.

Knott, J. 2010. Student learning gains in traditional and interactive engagement classrooms at a public, four-year university measured by the geoscience concept inventory (GCI). *Geological Society of America Abstracts With Programs* 42 (5): 555.

Kortz, K. 2010. Does peer discussion during use of clicker questions result in deep learning? *Geological Society of America Abstracts With Programs* 42 (5): 554.

Kortz, K., and J. Smay. 2010. *Lecture tutorials for introductory geosciences*. Nashville, TN: W.H. Freeman.

Libarkin, J. C., and S. W. Anderson. 2005. Assessment of learning in entry-level geoscience courses: Results from the geoscience concept inventory. *Journal of Geoscience Education* 53 (4): 394–401.

Libarkin, J. C., and S. W. Anderson. 2006. The geoscience concept inventory: Application of Rasch analysis to concept inventory development in higher education. In *Applications of Rasch measurement in science education*, ed. X. Liu and W. Boone, 45–73. Maple Grove, MN: JAM Publishers.

Libarkin, J., J. Kurdziel, and S. Anderson. 2007. College student conceptions of geological time and the disconnect between ordering and scale. *Journal Geoscience Education* 55 (5): 413–422.

Madison, B., and L. Steen. 2008. Evolution of numeracy and the National Numeracy Network. *Numeracy* 1 (1): 5.

Massingham, P., and T. Herrington. 2006. Does attendance matter? An examination of student attitudes, participation, performance and attendance. *Journal of University Teaching & Learning Practice* 3 (2): 82–103.

Mayer, R., A. Stull, K. DeLeeuw, K. Almeroth, B. Bimber, D. Chun, M. Bulger, J. Campbell, A. Knight, and H. Zhang. 2009. Clickers in college classrooms: Fostering learning with questioning methods in large lecture classes. *Contemporary Educational Psychology* 34: 51–57.

Mazur, E. 1997. *Peer instruction: A user's manual*. New York: Prentice Hall.

McConnell, D., D. Steer, and K. Owens. 2003. Assessment and active learning strategies for introductory geology courses. *Journal of Geoscience Education* 51 (2): 205–16.

National Research Council (NRC). 1997. *Science teaching reconsidered,* Washington, DC: National Academies Press.

National Research Council (NRC). 2000. *How people learn: Brain, mind, experience, and school. Expanded edition*. Washington, DC: National Academies Press.

National Science Foundation (NSF). 1996. *Shaping the future: New expectations for undergraduate education in science, mathematics, engineering, and technology*. Washington DC: NSF.

Neal, J. C., and M. A. Langer. 1992. A framework of teaching options for content area instruction: Mediated instruction of text. *Journal of Reading* 36 (3): 227–230.

Nichol, D. J., and J. T. Boyle. 2003. Peer instruction versus class-wide discussion in large classes: a comparison of two interaction methods in the wired classroom. *Studies in Higher Education* 28 (4): 457–473.

Roschelle, J., W. R. Penuel, and L. Abrahamson. 2004. The networked classroom. *Education Leadership* 61 (5): 50–54.

Semken, S. 2010. Factors that discourage persistent use of student personal response devices by earth and space faculty. *Geological Society of America Abstracts with Programs* 42 (5): 555.

Sibley, D. 2005. Visual abilities and misconceptions about plate tectonics. *Journal of Geoscience Education* 53: 471–477.

Steer, D. N., C. C. Knight, K. D. Owens, and D. A. McConnell. 2005. Challenging student ideas about Earth's interior structure using a model-based, conceptual change approach in a large class setting. *Journal of Geoscience Education* 53: 415–421.

Steer, D. N., D. A. McConnell, K. Gray, K. Kortz, and L. Liang. 2009. Analysis of student responses to peer-instruction conceptual questions answered using an electronic response system: Trends by gender and ethnicity. *The Science Educator* 18 (2): 30–38.

Symons, S., and M. Pressley. 1993. Prior knowledge affects text search success and extraction of information. *Reading Research Quarterly* 28 (3): 250–261.

Wren, K. 2011. Active teaching improves test scores and attendance compared to traditional lecture method. *www.aaas.org/news/releases/2011/0512sp_teaching.shtml*

Yeend, R., M. Loverude, and B. Gonzales. 2001. Student understanding of density: A cross-age investigation. Proceedings of the 2001 Physics Education Research Conference.

Zoback, M. 2001. Grand challenges in earth and environmental sciences: Science, stewardship, and service for the twenty-first century. *GSA Today* 11 (12): 41–47.

Revising Majors Biology: A Departmental Journey

Elizabeth Allan
University of Central Oklahoma

Setting

The University of Central Oklahoma (UCO) was founded in 1890 as the state's first public institution of higher learning. The main campus is located in the metropolitan Oklahoma City area and with over 17,000 students it is also the largest regional institution. Designated as a metropolitan university, UCO has 116 undergraduate majors and 55 graduate programs. UCO is ranked among the top universities nationally in residence life and as one of the top universities to work for by the *Chronicle of Higher Education*. The university has five colleges, a graduate school, and two programs (Academy of Contemporary Music and the Forensic Science Institute) that have specialized degree programs.

The UCO College of Mathematics and Science has a long history of providing an outstanding education in mathematics and science at the undergraduate and graduate levels with an emphasis on Transformative Learning (Toufic 2000; Hensel 2004; Barthell et al. 2010). The transformative learning initiative includes six areas of emphasis: (1) disciplinary knowledge; (2) leadership; (3) research, creative, and scholarly activities; (4) civic engagement and service learning; (5) global and cultural competencies; and (6) health and wellness. Several existing funding programs support one or more areas of transformative learning in CMS through mentor- and peer-based approaches to learning. An NSF-funded STEP grant, for example, supports undergraduate education through bridge programs that emphasize research experiences for its incoming students (both freshmen and community college transfers). The STEP@UCO program, with additional funding from Oklahoma EPSCoR, also incorporates high school educators (two per year for five years) from the Oklahoma City Metropolitan Area into the summer bridge program; an NSF S-STEM grant provides scholarships to students in the STEP@UCO cohorts to further enhance undergraduate retention in STEM disciplines. An NSF REU program has, for the last two years, targeted high school instructors for participation in an international research program that takes place in the Aegean region (Bulgaria, Greece, and Turkey).

The College has accredited programs in the departments of computer science, engineering, chemistry, nursing, and funeral science with the undergraduate science education program fully accredited by the National Council for Accreditation of Teacher Education (NCATE) as part of the college of education and professions studies unit and the National Science Teachers Association (NSTA).

The biology department is the largest in the university and has over 900 majors taught by 30 full-time faculty members. Three of the faculty members are administrators across the campus and two are shared with other departments. Biology faculty hold the greatest number of both internally and externally funded grants and produce the majority of publications. The majority of faculty members have research programs in which they involve undergraduates. Each faculty has a teaching load requirement of 12 semester hours per semester, unless they have been released from courses due to grants or other duties. All faculty members are assigned to committees within the department, college, and university and several serve in state and national organizations.

The Perfect Storm for Reform

The first course in the biology core curriculum sequence, Biology 1204, was developed in 2004 and initially taught in 2005 in response to poor student performance in upper-level biology courses. As a regional metropolitan institution, the UCO student population comes from a mixture of rural schools, urban inner-city schools, and suburban schools. As a result of differential course selection in high school, students were enrolling in courses with varying levels of both knowledge of and exposure to core scientific ideas. The Biology 1204 course was designed to bring all students up to the level of knowledge necessary for successful completion of subsequent core courses while introducing them to fundamental content not necessarily covered in a high school course. The course had a common syllabus with specific allocations of time on subjects, identical grading scale, exam schedule, and course policies.

When first developed, the course was indicative of most beginning biology courses (AAAS 2010). There was too much content and too few meaningful interactions with the curriculum. The textbook selected was a standard textbook with a high level of depth and breadth of the subjects, and while the faculty agreed upon a standard syllabus, there was little consistency among instructors. Standardized course content was developed, including learning objectives to guide student assessment. Students were generally bimodal in their success with a heavy concentration in the lower percentiles. The course quickly became one of the most failed courses in the university (mean failure rate < 50%).

Following the first year's implementation of the course, the department underwent a voluntary Council on Undergraduate Research (CUR) review beginning in 2006 with the review and findings presented in 2007. The most significant result of the CUR review was a revision of the core curriculum to provide a common depth and breadth of knowledge for all students. Several courses were developed and a clear progression through the curriculum was created using course prerequisites. Biology 1204 was retained as the entry-level course and became the "gatekeeper" for progression through the curriculum. Student success was essentially unchanged as the course was unchanged. However, student progression through the new core curriculum was dependent upon success in this course and faculty realized the need for a review of both what was taught and at what level of complexity.

As a result of national and state pressures and an impending accreditation visit by the Higher Learning Council, the university implemented a mandatory assessment system based on the central tenants of the university. The assessment system also required evidence of student performance on the learning objectives set forth by the department. Because of the timing of both the need for new courses and the mandatory development of an assessment system, assessments were built into the new courses as they were being developed. Assessments were also developed for existing courses, including Biology 1204. The result was the beginning of a comprehensive assessment system designed to measure the departmental learning objectives, including content knowledge, critical-thinking skills, problem-solving abilities, and scientific literacy.

These three events—the development of an introductory course with a low success rate; a CUR review that revealed the need for a stronger, more robust core; and the necessity of developing and implementing an assessment system—resulted in significant changes in the Biology 1204 curriculum and in the way in which faculty teach the course and consequentially, how students experience the course. Those changes and the process through which change occurred are described below.

Instructors of the Biology 1204 course are all full-time instructors or tenure track faculty members. After the 2008 CUR review, and in light of current research in undergraduate education instruction (Jones 1994; Bain 2004; PKAL 2007) the biology department made the decision to have permanent full-time faculty teach the core courses to ensure the consistency and quality of instruction. An additional factor in the decision was the availability for meetings of faculty who teach the course. The department decided to require all faculty who taught the course be involved in the design of the assessment of the course and any revisions of the curriculum. It is important to note that with a small number of faculty, a member of the Biology 1204 committee is more than likely a member of another course committee. This meant there was increased consistency between course development and revision across the course and contributed to the success of the Biology 1204 revision. One faculty member teaching the course was designated the chair of the course and set meeting times and agendas, guided discussions, and so on. Another significant task of each course committee was to begin the development of assessments within the class that contribute to the assessments of the departmental learning objectives.

The Biology 1204 committee met over the course of the 2009–2010 academic year to begin the initial review and revision of the curriculum. At each meeting the committee discussed a section of the course content and reviewed the student learning goals associated with the content. Each content area was reviewed for relevance, the learning objectives were reviewed for depth (e.g., Were they too shallow? Specific? Irrelevant?), and consensus was reached on the approximate amount of time to spend on the subject. Revisions to the course syllabus were made once all members of the committee were in agreement.

The committee spent significant time on reviewing this course's curriculum for its impact on future curriculum. As mentioned previously, several members of the committee were also teaching courses for which Biology 1204 was a prerequisite. This insight, along with the experience of having taught the course before, led relevance to the discussion. Often the conversations were about what knowledge students needed to be successful in a course, and what content was going to be covered in depth in a future course and therefore could have reduced coverage in this course.

Other times the conversation was remarkable insight into our teaching practices. Faculty often determined that they were spending significantly more (or less) time than others on content areas. This, in turn, led to further discussions of depth and breadth of content and the assignments we used to support the concepts. While awkward at first, eventually the faculty discussions were very productive with significant sharing of practices, ideas and opinions, both oppositional and supportive.

This interactive review of content and instructional practices had several implications for the faculty. Most faculty on the committee had little or no experience with collaboration around curriculum. Historically, individuals "owned" courses and the content in that course. Little was known about what was taught in each course and as a consequence, some content was covered multiple times and other content was not covered at all. This experience, and other committees' work on development and revision of other core courses, led to a greater awareness of the need for vertical and horizontal articulation.

In the process of reviewing the curriculum, it became clear that our students were not unlike those found in K–12 classrooms (NRC 2005). Our students come with preconceived notions of how the world works, and to be successful they must develop competency in learning about science. To do so, they must have the same three things that K–12 students must have to be successful. This is oftentimes in direct opposition to how instruction happens in a university lecture course. This dichotomy between what students need to learn and how instruction took place became evident as discussions about *what* to teach quickly became *how* to teach it. Frustration over student success often spilled over into blaming students for their inability or unwillingness to learn.

It was in the process of these discussions that the committee began to recognize a second, equally important goal of the course. While the main purpose of the course is to provide student access to fundamental biological content, a second goal became the development of students' cognitive and study skills. Our students were not dissimilar to other students and our curriculum and instruction problems are not dissimilar to others (Ulriksen, Madsen, and Holmegaard 2010; Abrams 2011; Friedman and Mandel 2011; Moses 2011). The majority of students in Biology 1204 were not prepared to take on the rigor and style of learning needed for success in a college-level biology major's course. This is significantly different from stating that our students were not *able* to be successful. Shifting from blaming students to one of understanding the needs of our students caused other significant changes in the course and instructor actions. It became as important to examine *what* the students did with the material as *how* and *what* was presented.

This ties back to the required assessment system, as the assessments became not just a requirement of the administration to meet accreditation demands, but an important part of the learning process. The Biology 1204 committee began to examine the assignments used in the class for their alignment to the learning objectives set forth by the biology department. The committee wanted to design assignments that both satisfy the assessment system requirements but also that build student skills. To date, three assignments have been developed. Instructors are collecting the data and sending it to the assessment committee who are in the process of analyzing the data.

The first assessment is a pretest/posttest. Students take a 20-question pretest the first week of class and again during their final exam. The pretest is representative of the breadth of content in the course and was approved by the committee. Student data is collected using student ID numbers and data are not separated out by instructor. The decision to look at the population as a whole rather than by section eliminated many of the concerns initially expressed by the instructors.

While the committee has not collected data long enough to be able to have sufficient data for programmatic use, the average increase in raw score per semester is 69%. The assessment committee is in the process of determining future plans for assessment, but one future goal is disaggregating the data by question to determine which questions (and therefore which content area) students are most and least successful on. This in turn will provide data for use in program improvement. Long-term data will also be collected so as to look for trends.

The second common assessment is a library assignment. The committee chose to introduce biology students to scientific literature early in their program, thereby addressing one of the department's top student-learning objectives. Students attend a class in the library where they receive instruction on how to use the library databases and then compare a scientific journal article to one on similar content published in a popular magazine. Initial data analyses indicate that students are very successful at completing the assignment (94.7% complete the assignment; 89.4% can search a database; 93.3% showed evidence in discerning the difference between a journal article and a popular magazine). Instructors use a common set of instructions and rubric for grading the assignment.

A third assessment is currently being used by more than half of the instructors and is being considered by the remainder. This assessment provides students with a journal article and has them answer questions about the content and scientific processes used in the article. This is a follow-up to their library assignment and extends their understanding about scientific literature while advancing their content knowledge.

These three assessments provide data about student performance but are not the only ways in which student learning needs are being addressed. Student fees are being used to fund supplemental instructors (SIs). These supplemental instructors are chosen from students who have successfully completed the course and are now enrolled in the next course in the curriculum sequence. SIs attend each class and hold two study sessions outside of the regular meeting time each week. Instructors can, and do, provide extra material for these sessions that provide something for students and the SI to interact with. No data is currently gathered on the success rate of students who attend SI sessions versus those that do not but anecdotal evidence indicates students believe they are benefitting from attending.

The biology department also provides free tutoring services for students. This year, some tutoring sessions are specifically marked for Biology 1204 and some of these sessions are held in the student dormitories. Students are free to attend any of the tutoring sessions available, but by specifying specific tutors for the course we hope they will gravitate toward those sessions and work together.

Finally, all faculty members require students to subscribe to the online course materials. These materials provide students with the opportunity for additional information, practice, and manipulation of the material through quizzes and tests. Various instructors make differing uses of the online materials but students are required to subscribe and use the website during the semester.

Individual instructors have other methods by which they attempt to increase student learning. Some provide examples from their personal research; others provide specific instruction in using textbooks and other supplemental materials; others use demonstrations and hands-on materials. Regardless of the specifics, instructors have become aware of the need to find ways that go beyond the traditional lecture to help students understand the content.

Next Steps

The Biology 1204 committee has several opportunities in the coming semesters. There is a proposal to open a tutoring center in the science building that will be staffed with science tutors. Discussion is occurring about building more assignments into the Biology 1204 course that require students to grapple with the content and requiring them to be completed in the tutoring sessions. The university has opened a center for excellence in teaching that will also be providing support for faculty who choose to improve their teaching skills.

Taken together, these are small steps toward addressing the cognitive and learning needs of our students. Much more needs to be done to move forward the change in perception that has begun in the department. Faculty continue to struggle to hold to high expectations while attempting to modify instructional techniques to address our students' learning needs

Several years ago it would not have been unusual to hear biology faculty talk about what content needs to be taught in a course. Content knowledge is a comfortable space for higher education faculty. It would have been much more unlikely to hear biology faculty teaching a major's course discuss pedagogy. With the advent of a mandatory assessment system, the redesign of the core curriculum, and a growing awareness of the cognitive needs of our students, attitudes and practices are changing. And change is coming from K–12 classrooms. With the publication of *Framework for K–12 Science Education* (NRC 2012) and the recent release of the *Next Generation Science Standards* (2013), K–12 science education has a significant conceptual shift coming. As students emerge from K–12 classrooms, where they have experienced curriculum that integrates scientific practices, crosscutting concepts, and disciplinary core ideas, they will place unique and different demands on university faculty. Higher education faculty will be teaching students that have a contextual understanding with regard to the content of science, the processes through which that content is acquired, and the application of that content. Faculty that maintain their "silo" of content knowledge without regard to the integration of these three dimensions have the potential to be perceived by students as nonexperts or at the least not knowledgeable. If students have developed a deep understanding and application of content by curriculum that builds coherently across the K–12 years, they will expect the same in college courses across the disciplines and through their program of study. This may, in turn, provide a dilemma for faculty who are required in their research to become more and more specialized. Regardless, changes in the K–12 arena will necessitate changes in higher education. Faculty in departments that develop the ability to review and revise curriculum will be better poised to handle these changes.

References

Abrams, W. P. 2011. Mathematical vignettes in a university-controlled freshman seminar, or "oh—that's cool!" *PRIMUS* 21 (3): 274–282.

Achieve Inc. 2013. Next generation science standards. *www.nextgenscience.org/next-generation-science-standards*

American Association for the Advancement of Science (AAAS). 2010. *Vision and change: A call to action.* Washington, DC: AAAS. *http://visionandchange.org/files/2011/03/Revised-Vision-and-Change-Final-Report.pdf*

Bain, K. 2004. *What the best college teachers do.* Cambridge, MA: Harvard University Press.

Barthell, J., E. Cunliff, K. Gage, W. Radke, and C. Steele. 2010. Transformative learning—Collaborating to enhance student learning. Proceedings of the 115th annual meeting of NCA/The Higher Learning Commission.

Friedman, B. A., and R. G. Mandel. 2011. Motivation predictors of college student academic performance and retention. *Journal of College Student Retention Research Theory and Practice* 13 (1): 1–15.

Hensel, N. 2004. From the Executive Officer. *Council on Undergraduate Research Quarterly* 25: 53.

Jones, R. 1994. First-year students: Their only year? *Journal of College Science Teaching* 23 (6): 35–40.

Moses, L. 2011. Are math readiness and personality predictive of first-year retention in engineering? The *Journal of Psychology* 145 (3): 229–245.

National Research Council (NRC). 2005. *How students learn: Science in the classroom.* Washington, DC: National Academies Press.

National Research Council (NRC). 2012. *A framework for K–12 science education: Practices, crosscutting concepts, and core ideas.* Washington, DC: National Academies Press.

Project Kaleidoscope (PKAL). 2007. *What works: Building natural science communities.* Washington, DC: PKAL.

Toufic, H. M. 2000. *At the interface of scholarship and teaching: How to develop and administer institutional undergraduate research programs.* Washington, DC: Council on Undergraduate Research

Ulriksen, L., L. M. Madsen, and H. T. Holmegaard. 2010. What do we know about explanations for drop out/opt out among young people from STM higher education programs? *Studies in Science Education* 46 (2): 209–244.

Students Teaching Students: Jigsawing Through an Environmental Biology Course

Thomas R. Lord
Indiana University of Pennsylvania

Setting

The environmental biology course discussed in this chapter was designed for non-science majors, most of whom were preservice elementary education students. The course was taught by faculty of the biology department at Indiana University of Pennsylvania, and follows the recent recommendations put forth by the American Association for the Advancement of Science (AAAS), the National Research Council (NRC) and the National Science Teachers Association (NSTA). Instruction in the course is designed around the constructivist teaching model based on inquiry instruction and features the student-teaching-student design of instruction.

Students have been evaluated on their learning in science courses for decades. The most favored method of gauging learning is through instructor-generated exams that are aimed at what the class members can recall from their readings and lecture materials. Assessing what students can recall from the professor's presentations generally measures the facts, definitions, and terms recalled from class and not to what the test takers can apply or relate to they have not heard in the class. For this to occur, it's believed students need to take a larger role in their learning; and this has led educational leaders to propose involving students more actively in the lesson rather than treating them as passive spectators (Knight and Wood 2005; Walker et al. 2008; Wood 2009).

To discover how to engage class members more in their learning, the American Association for the Advancement of Science (AAAS) invited a number of college science instructors, recognized as leaders and innovators in the field, to a weeklong meeting to discuss how this could be accomplished. Also invited to attend the sessions were college biology student-leaders whose input was sought by the organizers in order to get the learners' prospective. The attendees not only examined what was being covered in the science courses generally taught during the typical eight semesters of education, but how the professors were teaching the subject matter to the students. To enable this huge task to be accomplished in the short time provided during the week, the leaders decided to direct their efforts to instruction in the life sciences.

Teaching Through Inquiry Instruction

The participants also discussed several innovative instructional methods that had been shown to be effective with young adults. The most frequently talked about strategy discussed by the conference participants was the innovative use of inquiry in the learning process. Inquiry instruction involves experiencing content through questions that students attempt to answer by sharing and discussing their thinking with teammates. Contemporary learning theorists believe that as team members consider the suggestions and responses of their partners for the challenges, they blend the new information with their preconceived understandings about the topic and build new knowledge for themselves (Fuller 2002). One of the major components of the inquiry model, therefore, is that students can effectively learn course content from each other. The role of the professor in the inquiry strategy shifts from being the presenter of content to be learned by students to being the manager of a student's learning the content (Fosnot 1996; Gafney and Varma-Nelson 2007).

A good example of how inquiry works in biology is in the instruction of cell structure and function to introductory life science students. Traditionally, class members are taught the various components of the cell via the instructor pointing to a drawing of each cellular element on an overhead transparency or a PowerPoint slide and describing its makeup and function. As the instructor goes over the cell membrane in this traditional way, for example, class members quickly sketch the structure in their notebooks and jot down its various components. In the meantime, the professor has likely moved on to a description of passive and active diffusion and various structures within the membrane itself.

An inquiry instructor, on the other hand, would likely follow the 5E Instructional Model (Bybee 1997) by introducing the lesson with a discrepant event, attention-grabbing reading, a short demonstration, or a video clip on diffusion. The instructor could, for example, pass around several inflated balloons, each containing the concentrate of a different spice (e.g., cinnamon or vanilla) and then ask students in the class to describe the smell of each balloon. Known as an *engage*, the event sets up the topic of the lesson for the students. Class members, working in teams of 3–5, are next challenged to come up with (*explore*) 10 factors that would influence the movement through the membrane of the molecules just sensed. Jotting down their answers on a sheet of paper, teams are given several minutes to come up with their ideas before their answer sheet is collected. Once all the answers are retrieved, the instructor randomly selects a response and asks one of the writers of the sheet to share the team's ideas with the rest of the class (*explain*). When the shared answer is completed and verified, the professor randomly chooses a second paper from the stack and asks a member of that group to share one of that team's answers. The procedure is continued as students from various teams share (or teach) their answers to their classmates. Eventually the instructor displays all the potential answers on the PowerPoint screen to round out the discussion (*elaborate*; the final "E" is *evaluate*). In the inquiry model, when students teach and learn from other students in such a manner, the strategy is known as a jigsaw (Figure 11.1).

Figure 11.1. Schematic of a Jigsaw for a Challenge Question Asked to the Teams by the Instructor and a Response Given by One of a Team's Members

In this jigsaw method, students in small teams discuss a challenge question posed by the instructor and together, come to a consensus on the issue (each team is challenged to answer the same question). To hold them accountable, the teams jot down their answers on a sheet of paper and give it to the instructor.

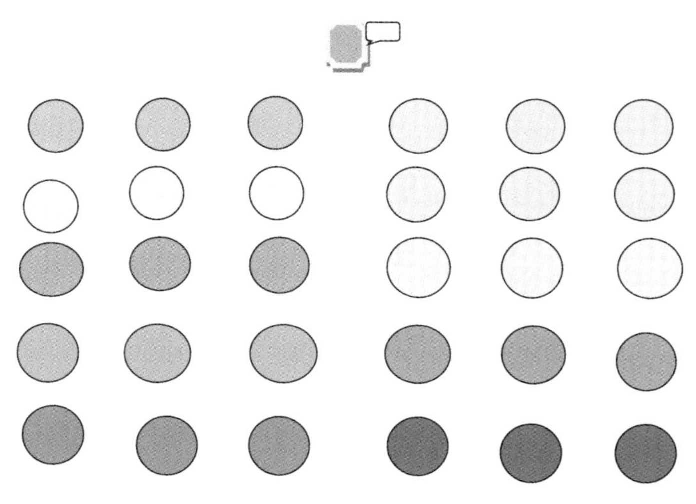

Once the instructor obtains all the teams' responses, one of the group's answer sheets is drawn from the stack by the professor, who asks one of the members on the team who created the sheet to share their answer with the rest of the class. The student sharing routine is continued until completed.

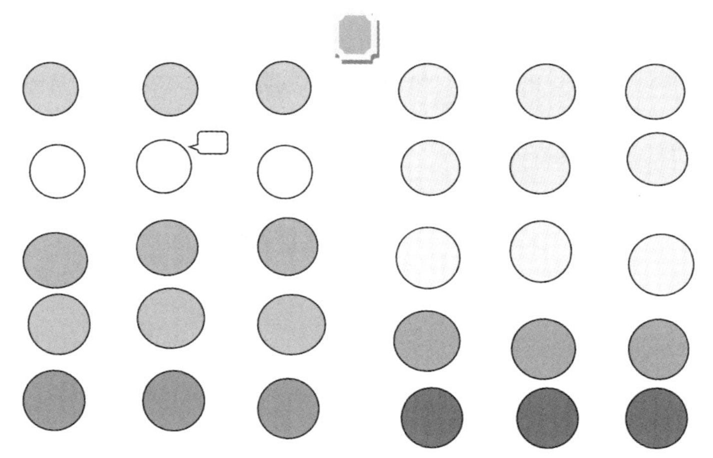

Working Models of Jigsaw

There are several teaching techniques that employ alternative jigsaw models. In my environmental biology class, student teams of three to four members select a course issue from a series of topics on the first day of class that they will research. To help guide the researching groups, each issue contains a list of items that team members must include in their presentation. For example, a team that selects the topic *environmental succession* is required to include the various biotic and abiotic characteristics of each successional phase in their report. Team members of this topic refine their teaching of the topic by taking one the serial phases of succession as they research the characteristics of that one phase. One participant, for example, would research the characteristics of pond succession, while a partner team member would investigate the succession of marshes; the pattern is continued as a third team participant researches the development of an old field, and a fourth member of the team examines the succession of a young forest to its climax. Many times, the student group with this topic will choose to teach the class what they have researched with short PowerPoint presentations of 10–15 minutes (Figure 11.2).

Figure 11.2. Jigsaw With Group Presentations.

In this jigsaw method, each team selects one of the topics in environmental biology to present to their colleagues in the class. The presenting team members research their topics, collect graphics to support their talk, and prepare PowerPoint slides for the presentation.

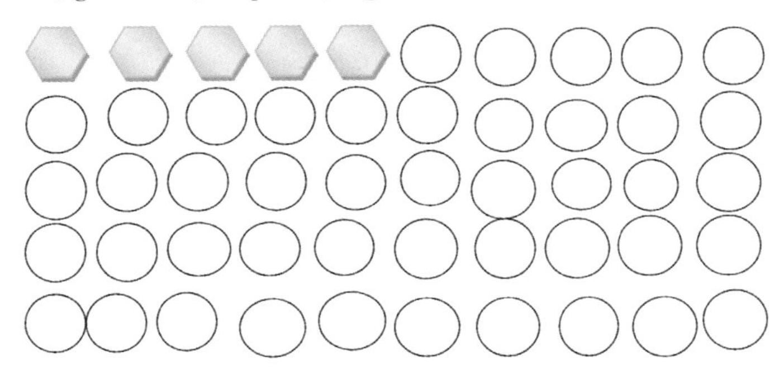

On the day of the presentation, presenters stand behind a podium and share on the PowerPoint slides the information they discovered through the research in 10-minute snippets. Students in the class may ask questions or inject topical comments throughout the presentation.

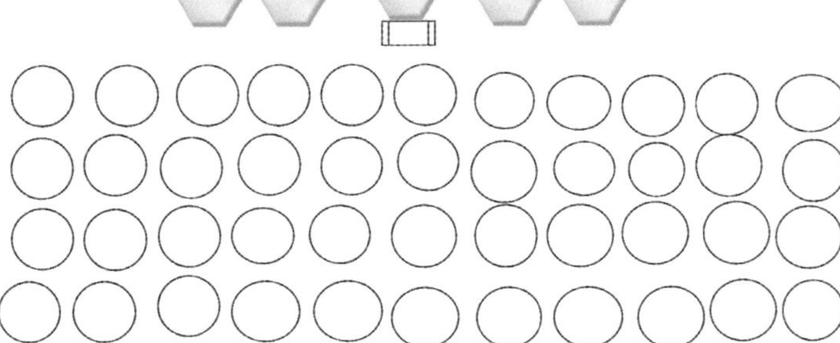

Besides presenting information to the class in my environmental biology course, I require teams to produce a multichapter term paper on what the group has discovered. Each member of the team is responsible for one chapter of the document that includes what he or she has researched and presented. To assure that chapters are proportionally the same length, students are told they cannot submit a chapter shorter than four or longer than five pages. The creators of the term paper will be given two grades for their efforts, one for the chapter they have written and a second grade for the overall composition of the document. Members are encouraged to proofread the entire document (his or her own and other teammates' writing) for spelling, punctuation, and grammatical errors.

The jigsaw technique can be used just as effectively with small clusters of students instead of groups (clusters are nonpresenting members in the class that have been divided into four or six groups). In the environmental biology class, I initially follow a similar format to the one above. I divide teams into clusters of five students and present them with a topic of which each member in each team is responsible for a different aspect. In a class studying North American biomes, for example, group members are required to include the characteristic flora and fauna, along with several important abiotic factors, of each biome region (deserts, temperate forests, grasslands, taiga, and arctic). Team members are also encouraged to find photographs, maps, and other graphics of their selected biome to share with the groups they are teaching. Team members are told they are required to complete their presentation in a 10-minute period so as to not interfere with a teammate's presentation that will take place immediately following his or her talk. In this example in environmental biology, teams are composed of five members, each with a different biome to share. On the day of the presentation, each researcher presents to class members in his or her cluster what they have found out about their biome. As long as the 10-minute time allotment is followed during each rotation from cluster to cluster, the entire class will have learned the information about a biome from the five presenters (Figure 11.3, p. 150). As in the previous example, the term paper required by the group would contain four to five chapters, each contributed by one member of the team.

I use jigsaw instruction with my environmental biology class in other ways. Instead of assigning various members of a group to research a different biome, I've assigned all the members of a team to research all aspects of the same biome. To make the jigsaw work, I assign the members of a second team in the class to research the characteristics of a second North American biome; students in a third team are assigned to investigate a third biome, students on a fourth team are assigned to explore a fourth biome, and members of a fifth team are assigned to research the fifth biome. In this technique, since members on a single team will research the same biome, they must each select a different characteristic of their biome to write about in their term paper.

On the day of the presentation, one member from each of the five researching teams will join one member from each of the other teams. This brings together one individual in each team who has researched a particular biome. The five researchers at each cluster is then given 10 minutes to present what he or she has learned about his or her particular biome (Figure 11.4, p. 151).

Another way that I use the jigsaw technique in environmental biology class is to employ the debate technique often used in traditional classes. In the jigsaw model of the debate strategy, two teams of three or four students are asked to choose either to support or contest a contemporary issue. One of the topics that go well with the debate strategy is climate change. Members of one team research their side of the topic and organize a plan to use in a classroom debate against a second team who has investigated the alternate side to the topic. Research papers produced by each team consist of three to four chapters, each written by one member of the team. The topics of each chapter must be different and be supported by scientific documentation. For example, teams arguing that climate change will not be as devastating as many believe must back their stance with reliable evidence. This group, for instance, could show evidence that Earth's warming could open waterways through the Arctic, which will be good for commerce and trade. The opposing team would likely counter that opening the ice sheet in the arctic will likely lead to the extinction of the polar bear. On the day of the debate the two groups will be

Figure 11.3. Jigsaw Schematic of a Team of Five Members Presenting Different Aspects of a Specific Topic to Their Classmates

After all the groups have selected a topic, each team member on one of the teams researches a particular segment of the topic to present to their classmates.

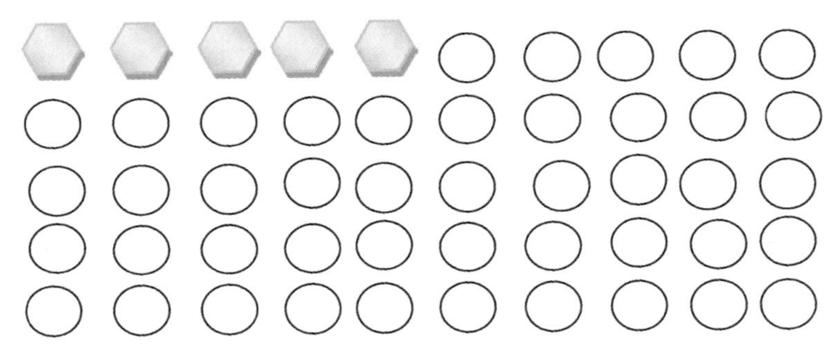

On the day of their presentation, one member from the team will join a cluster of their classmates and present what they have discovered about the topic. After 10 minutes, the presenter will move to the next cluster of classmates and repeat the presentation to the second group. This routine is repeated until the whole class has heard the presentation.

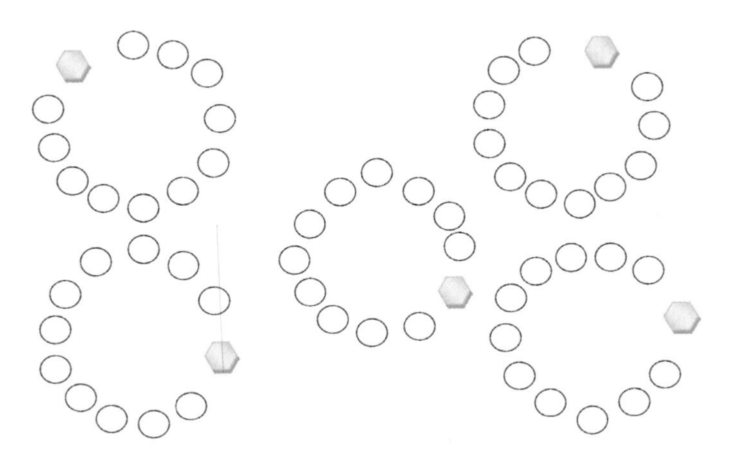

Figure 11.4. Jigsaw Schematic of Five-Member Teams; All Members of a Team Learn a Specific Aspect of a Topic to Share With Their Classmates

After all the groups have selected a topic of the course, each team member on one of the teams researches a particular segment of the topic to present to their classmates.

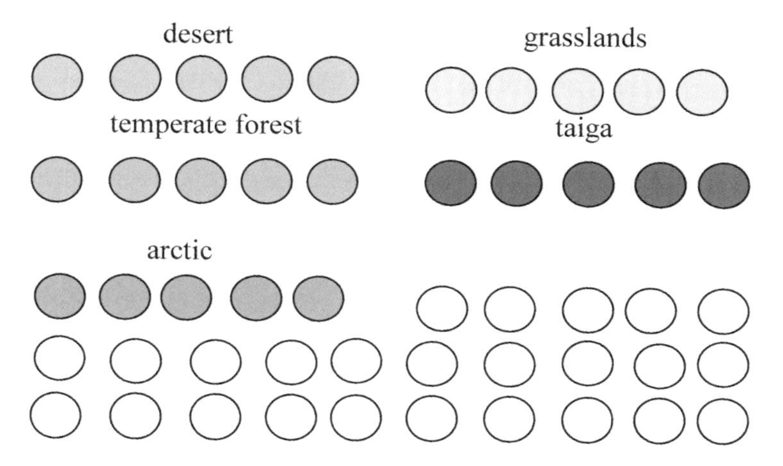

On the day of their presentation, one member from the team will join a cluster of their classmates and present what they have discovered about the topic from their research. After 10 minutes, the presenter will move to the next cluster of classmates and repeat what they research to the second group. This routine is repeated a third, fourth and fifth time until the whole class has heard the topic.

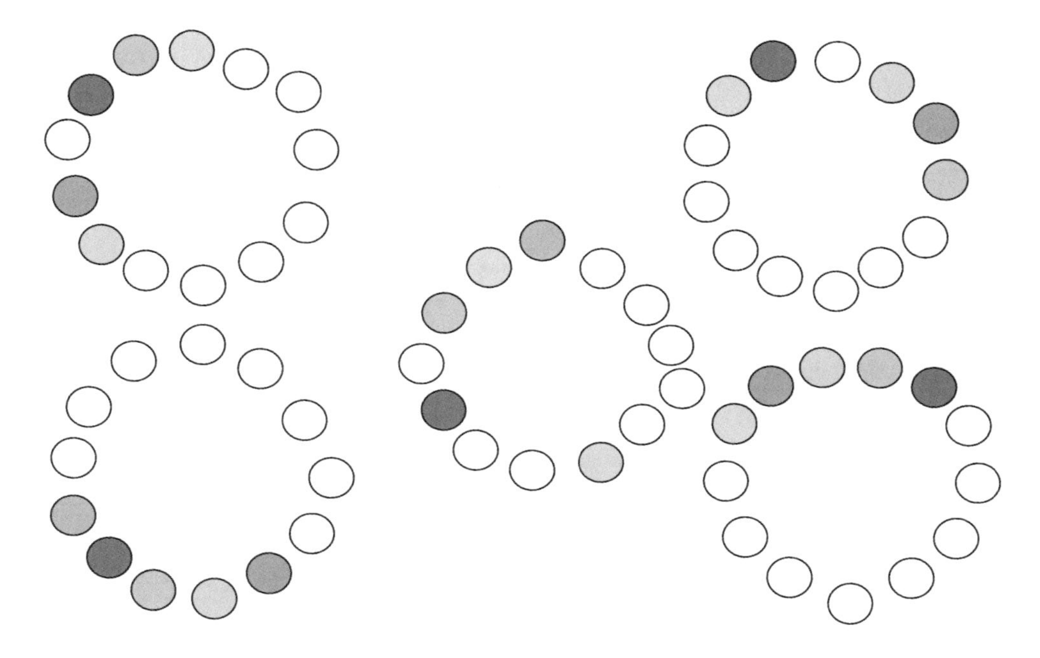

seated on one side or the other in front of the class. The teams will have equal time to present their side and to question each other. The course instructor will moderate the debate. At the conclusion of the class, the audience will select the winning team, justifying their decision with supportive reasons (Figure 11.5).

The debate format can be modified into a small-group format instead of the large group. Teams of two or three students work together to present in a "discourse" fashion the conflicting issues pertinent to topics in environmental biology (e.g., pros and cons of using biofuels on the growth and development of living things). The student group is then matched against two or three additional classmates who have researched the opposing stance on the same issue (Figure 11.6, p. 154). When the class meets for the discourse session on the topic, each student presents his or her side of the issue to a cluster of classmates. Halfway through the period, the presenting students move to a cluster that has not learned about that side of the issue, and present their side of the issue a second time. At the end of the class hour, every class member has learned both sides of the issue. Working with their discourse peers, the clustered students assess the quality of the pro and con presentations. This instructional method empowers students in the cluster to share feelings about the issues and reveal why they think one presenter did a better job than the other.

The final use of the jigsaw model used in my environmental biology course is what most often is thought of as learning through practical examples. Recommended for implementation for longer laboratory periods, this teaching scheme involves small groups of students researching separate course techniques and teaching what they've learned to their peers in the class. When, for example, the class is learning about air pollution control systems, five members of each team research the control procedures of just one of five categories of noxious pollutants (acids, gases, mercury, nitric oxides particulates, and volatile organic compounds). On the day the topic is discussed, one member of each team joins with one member of each of the other teams and the new group moves from one demonstration site in the room to another in an orderly fashion. When the group arrives at each site, the student who is knowledgeable about the process at that location teaches his companions about the cleansing procedure setup at that location. By moving to each station every 20–25 minutes, the entire group learns about the cleanup of a noxious pollutant at each site. This results in an understanding of pollution control by the entire class (Figure 11.7, p. 155).

Student Thoughts and Comments on Jigsawing

Overall, class members in the environmental biology class reacted positively to the student-teaching-student methods. Initially, the biggest concern class members taught by the various jigsaw methods have is whether they will learn the same level of content as they would if the professor taught them the lesson. This apprehension occurs despite research that has found the students learn as much, and in some cases more, when they teach one another (Cross 1990; Fuller 2002; Wood 2009). The anxiety can be further reduced when the class members realize that each presenting student is provided with a list of items they are required to include in their presentations and, further, that the student presenters go over their lesson with the professor before their talks. Typical comments are (text continued on p. 156):

- I like learning from other students because they explain things in a language I understand.

Figure 11.5. Jigsaw Schematic of the Student-Teaching-Student Debate Format

In this jigsaw method, two groups of three or four members select opposing stances of a contemporary issue. The members of each team research articles, texts, or documents to support their stance on the issue.

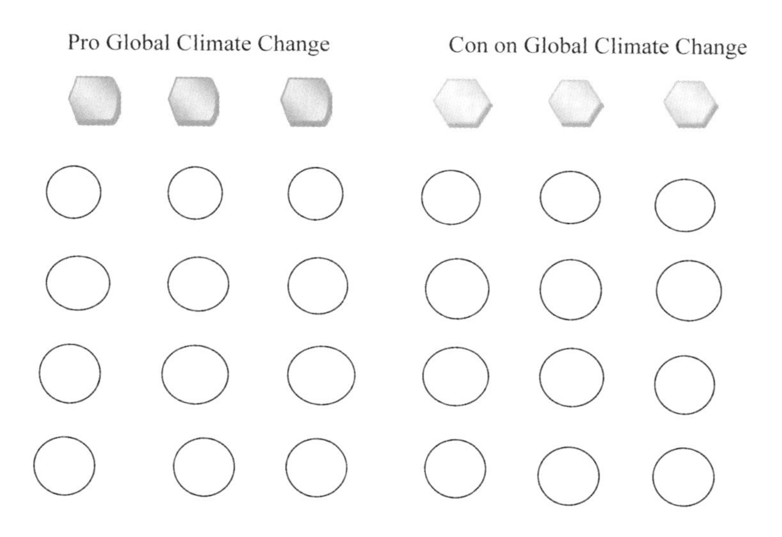

On the day of the debate, the participants arrange themselves in seats on either side of the classroom in front of their classmates. The moderator (course instructor) sits between the two groups and asks questions.

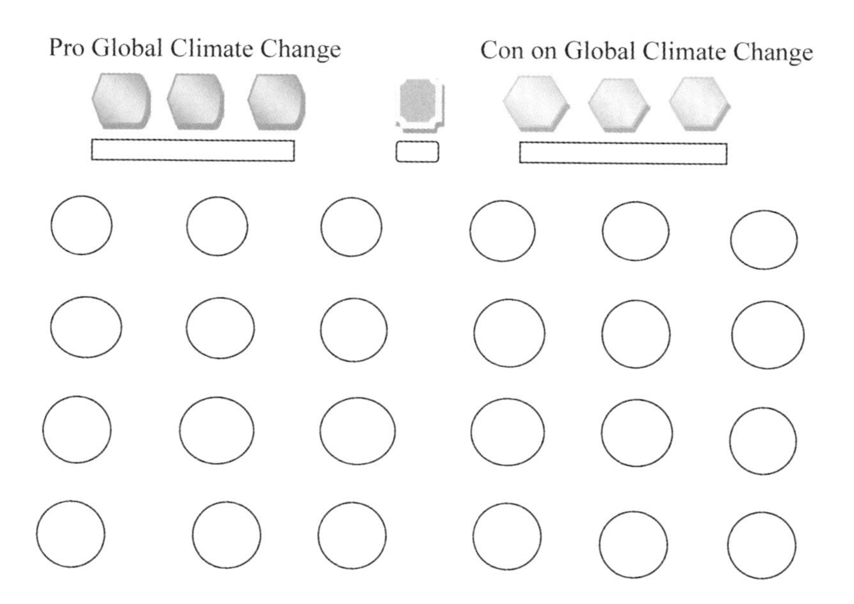

Figure 11.6. Jigsaw Schematic of a Discourse Between Two 2-Member Teams

The diagram below represent two teams, one that selected to research the favorable side of an issue (e.g., pro pesticide usage) and the other team selected to research the opposing side of the issue (con pesticide usage).

Each team researches their stance on the issue (together or separately), meeting periodically to share what they have discovered and to design graphic support items. On the day their issue will be presented, the two team members on each issue have identical documents to present to their classmates.

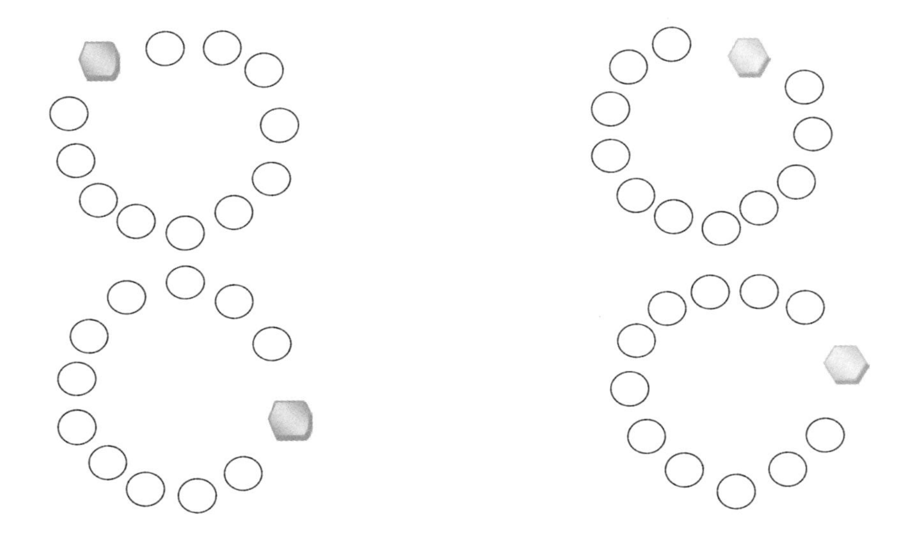

After 20–25 minutes, the instructor terminates the discussion and switches the presenters to a cluster that has not heard the other side of the issue. The presenters repeat their presentation to the second cluster. After a second time period of 20–25 minutes, the instructor closes the discussion and directs the four presenters to leave the room for 5 minutes.

Figure 11.7. Jigsaw Schematic of Teams of Five Members, All Learning Through Hands-On Course Techniques

After all the groups have selected a pollution clean-up strategy, all the members of each team research that technique and teach it to the members of other teams.

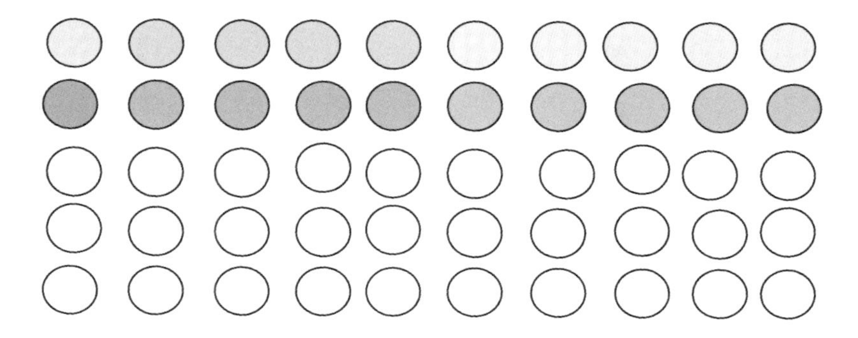

On the day of their presentation, one member from each team will join a cluster that is not represented by a knowledgeable person and talk about what he or she has researched. After 10 minutes, the teaching students will move to the next cluster of classmates and repeat what they researched to the second group. This routine is repeated a third, fourth, and fifth time.

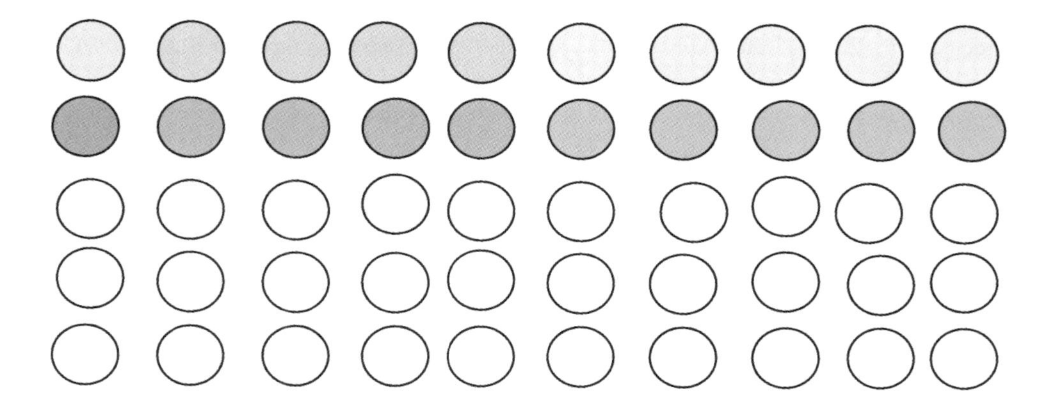

- I find it's less intimidating being taught by our friends than from the professor.
- When I learn from a friend the atmosphere in the class is more relaxed and fun.
- I am more likely to ask a question or share a concern with others when being taught by my friends in a small cluster instead of a professor in a large lecture hall.

One student even remarked, "I've always hated science and never learned anything from the science classes I've taken; however after taking environmental biology, I realize how interesting and fun science can be."

There are some students, however, who dislike the student-teaching-student scheme of learning. Opposing comments included:

- I learn better when the professor tells us in his own words what we should learn.
- I like being taught more from the professor when I can sit back and learn the way I'm use to.
- The jigsaw method seems more appropriate for high school students than college classes.

However, not only do most students enjoy teaching and learning from each other but studies show that they learn as much or more than they do from instructor-dominated classes. Research (e.g., Angelo and Cross 1993; Cross 1990; Huba and Freed 2000) has found that students taught through student-centered inquiry score higher on their exams and retain the information longer than students in non-hands-on classes. In a study by Handelsman and colleagues (2007) it was found that enhanced student activity heightens the thinking levels of the participants and leads to greater understanding and recall of what is being taught. Additionally, research by Ebert-May and colleagues (2003) concluded that students actively involved in the teaching and learning of their courses demonstrated significantly higher proficiency in the courses they were taking than students in traditional lecture courses. The findings of the students teaching each other in environmental biology support these report results. The overall majority of students in the environmental science course said they would like to take another science course taught through inquiry.

References

Angelo. T., and K. P. Cross. 1993. *Classroom assessment techniques: A handbook for college teachers*. San Francisco: Jossey Bass.

Bybee, R. W. 1997. *Achieving scientific literacy: From purpose to practice*. Portsmouth, NH: Heinemann.

Cross, K. P. 1990. Teaching to improve learning. *Journal of Excellence in College Teaching* 1: 9–22.

Ebert-May D., J. Bazli, and H. Lim. 2003. Disciplinary research strategies for assessment of learning. *BioScience* 53: 1221–1228.

Fosnot, C. 1996. *Constructivism: Theory, perspective and practice*. New York: Teachers College Press.

Fuller, R. 2002. *A love of discovery: Science education.* Dordrecht, Netherlands: Springer.

Gafney, L., and P. Varma-Nelson. 2007. Peer-led team learning. *Journal of Chemical Education* 84: 535–539.

Handelsman, J., S. Miller, and C. Pfund. 2007. *Scientific teaching.* New York: W.H. Freeman.

Huba, M., and J. Freed. 2000. *Learning-centered assessment on college campuses: Shifting the focus from teaching to learning*. Needham Heights, MA: Allyn and Bacon.

Knight, J., and W. Wood. 2005. Teaching more by lecturing less. *Cell Biology Education* 4: 298–310.

Walker, J., S. Cotner, P. Baepler, and M. Decker. 2008. A delicate balance: Integrating active learning into a large lecture course. *Life Science Education* 7: 361–367.

Wood, W. 2009. Innovations in teaching undergraduate biology and why we need them. *Annual Review of Developmental Biology* 25: 93–112.

The Great Debate: Revising an Old Strategy With New Frameworks

Teddie Phillipson-Mower
University of Louisville

Introduction

The development of critical thinking skills is imperative for democracy and the future of our country. Therefore, continual revisiting of the main elements and application of these skills throughout one's education is a necessity. Science, with its very framework embedded in a matrix of inquiry and logic, is the ideal discipline to practice and develop these skills while tapping into students' natural curiosity and interests. However, the reality is that undergraduate introductory science courses seldom move beyond the lecture. Seldom are the students asked to synthesize the information they receive and determine why and how it counts as knowledge. Strategies, such as the great debate, encourage and extend student thinking to understand and use science in everyday life decision making. This is especially important in the physical sciences where the basic science concepts and relevancy can be lost in the seemingly abstract and cognitive demands of mathematics.

There are those who would disagree that debates should involve novices—that it is for experts only. There are also those that would argue that there isn't time to cover all of the material (as opposed to teaching it). I would counter that the debates that didn't happen in undergraduate science education are now taking place in the public sphere without the same opportunity for science framework building and guidance in its use. Connecting science content with policy, local issues, and social contexts introduces students to thoughtful decision making and citizenship skills in the very complex system of 21st-century America. If this isn't a compelling reason, these are the people that will determine the future of science and education funding both publicly and privately.

Setting

The context for the learning project that will be discussed here is within an environmental education program that was developed to respond to the needs of formal and informal educators for advanced training in content and pedagogy that leads to environmental literacy. The setting is a large (<22,000) urban public research institution that grants two-year, four-year, and graduate degrees.

Housed in the college of education, the program also includes undergraduate honors students under a different course number. Most of the honors students are from the science disciplines. The program consists of four sequenced courses: (1) introduction to environmental education, (2) environmental education teaching methods, (3) a content course, and (4) environmental education capstone. Three of the courses are taught in a two-year rotation by the author. The third course, the content course, is chosen by the student and advisor from several cross campus possibilities. Field studies, instructional procedures in science at the zoo, and global climate change are three of the choices that are encouraged. The program was developed to meet the criteria of state teaching certification (endorsement on licensure) as well as being situated as the emphasis of study in a master's degree in teaching leadership. However, many of the students who come from other disciplines, especially the natural sciences, take the course as a way to develop skills that are becoming essential in landing their first jobs: the ability to communicate science to the general public.

Class sizes for the three core requirements have been between a low of 7 and a high of 21. Approximately one-fourth in the program are high school science teachers and a little over one-fourth elementary teachers. The rest are teachers from other disciplines (social studies, English, math, and the arts); informal teachers (working in nature centers, forest preserves, and county and local park systems); and honors students from geology, biology, and the physical sciences. The students, in general, have a high level of awareness about environmental issues and pro-environmental attitudes. They have a desire to be able to apply their learning to educating others about the environment.

Several lessons, activities, projects, and "teachable moments" help to support the emphasis on critical thinking in the program and many are revisited for reinforcement at various times in the program. For example, during the capstone course students find clients that need environmental programming and together choose one client for which to develop, implement, and assess a program. Outcomes for the program (from the framework in the syllabus) include teaching critical-thinking skills. For this, they are able to go back to the prior three courses for ideas on how to teach and assess this component. In the introductory to environmental education course, students are introduced to and develop their own critical thinking, which in this case is considered content knowledge (CK). In the environmental teaching methods course, students learn how to teach and assess critical thinking. This is known as pedagogical content knowledge (PCK) since it involves the professional knowledge of how best to teach the content of critical thinking (Shulman 1987). In the capstone course they further develop their PCK as they apply their knowledge within a context of active teaching and learning.

For the purposes of this chapter, I will share one component that is used to emphasize critical thinking and argumentation skills, what I call the great debate. This learning project was developed out of frustration that resulted from the lack of deep content exploration when using the strategy of debate in the science/environmental classroom. It appeared that critical thinking skills limited by low-level epistemological stances hampered a meaningful outcome.

Overview of the Program

The great debate meets all four of the NSES goals of school science:

- experience the richness and excitement of knowing about and understanding the natural world;

- use appropriate scientific processes and principles in making personal decisions;
- engage intelligently in public discourse and debate about matters of scientific and technological concern; and
- increase their economic productivity through the use of the knowledge, understanding, and skills of the scientifically literate person in their careers.

The NSES explicitly states that there should be more emphasis to promote inquiry through "science as argument and explanation," "using evidence and strategies for developing and revising an explanation," "groups of students often analyzing and synthesizing data after defending conclusions," and "applying the results of experiments to scientific arguments and explanations." In addition, changes in emphasis should include "learning subject matter disciplines in the context of inquiry, technology, science in personal and social perspectives, and history and nature of science" (NRC 1996, p. 113).

The great debate actively engages students at the university level in learning science as well as learning what science is, what it is not, and the assumptions inherent in the endeavor (nature of science). As pointed out in reform documents, "learning science is something students do, not something that is done to them" (NRC 1996). These types of experiences are essential (but not sufficient) to this particular population, environmental educators, as they will in turn teach the way they are taught. If we are to change science education goals from transmission to transformation, from an accumulation of facts to the construction of ideas, we must begin with the training of our future teachers (Erduran, Ardac, and Yamaci-Guzel 2006; Topcu, Sadler, and Yilmaz-Tuzan 2010). We must make this training explicit through modeling, reflection and supporting the development of their PCK.

Major Features of the Instructional Program

All one has to do is open Facebook, turn on the television, or pick up a newspaper to see misguided and thoughtless claims being made. Our political leaders continually show their lack of understanding of the nature of science by making statements such as, "evolution is just a theory" (as if a theory is equivalent to an idea), that "creation science should have equal time in the science classroom" (as if science is democratic and creation meets the definition of science), or that "decisions can't be made until the science is certain" (as if the tentative nature of science is a weakness instead of a strength). How many times have you heard, "well, everyone is entitled to their opinions" without evidence and justification to go with it? Our future as a democracy is dependent upon a citizenry that is capable of making decisions about a wide range of issues that are likely to intersect with science (NRC 1996; Topcu, Sadler, and Yilmaz-Tuzan 2010). We seem to be failing to convey the very central commitment of science, the necessity and power of evidence as the basis for knowing, decision making, and resolving conflict (Erduran, Ardac, and Yamaci-Guzel 2006; Osborne 2010).

Osborne (2010) points out that at a time when the ability to critically think is much needed and there is an economic necessity to maintain a scientific and technology base, students are losing interest in science. Under NCLB, testing students' abilities to recall snippets of information from a broad received body of knowledge has minimized the opportunity to do science in "the spirit of open inquiry and invention that it has fostered." If students are empowered to engage in questioning; determine what and how to collect data to answer their questions; discuss the limits and strengths of evidence collected; reflect on values, predispositions, and background as the source of bias, creativity, and imagination; and reason, justify, and deliberate conclusions and claims, they might find relevancy and have a renewed interest in science.

Argumentation refers to the interdisciplinary process of justifying claims and drawing conclusions using evidence and logical reasoning. Over the past few decades, science education research has analyzed argumentation discourse in the classroom, which has highlighted its importance in developing scientific literacy (Erduran, Simon, and Osborne 2004; Kelly 2007). Research in argumentation has been justified on philosophical, cognitive, and sociocultural grounds (Erduran, Simon, and Osborne 2004). Current philosophy of science perspectives emphasize the construction of knowledge through tentative explanations that are open to challenge and reinterpretation. Science as a way of knowing is based on the justification, validity, and consistency of claims through the evaluation and interpretation of indirect and direct evidence. The very heart of science is argumentation. Cognitive foundations of justification include the opportunity for children to externalize internal ideas of the natural world and to test them against the social dimensions of knowledge, beliefs, and values. This allows for development of the understanding of the relationship between claims and warrants allowing them to move to higher-order thinking that requires warranted claims. Socioscientific frameworks look at argumentation as the door to scientific discourse and the community of practice. There is justification in studying this discourse if it is significant to science teaching and learning (Erduran, Simon, and Osborne 2004).

Much of the research and teaching practice in scientific argumentation is focused on structured or semistructured instructional models that engage students in addressing what we know in science, what evidence supports it, and what counts as evidence (e.g., Proulx 2004; Sampson and Grooms 2010; Sampson and Grooms 2009). This is understandable, given that the contexts are science courses and science content learning is the main focus. However, as the reforms point out, it is also important to consider everyday science decision making as well (NRC 1996, pp. 107–108).

In his book on humanistic science education, Aikenhead (2006) demonstrates that the literature on decision making in socioscientific contexts is consistent in suggesting that values, personal experience, and common sense hold much higher priority than evidence, scientific information, and understanding of the nature of science. When students are emotionally involved in the issues, values trump science unless they (the students) anticipate participation in an action-oriented event such as finding a solution to a local or personal issue. Students are limited in making decisions both by their understanding of science content and their naïve perceptions of evidence. There is also difficulty in making clear distinctions between "frontier science" and "core science." Such misunderstanding and using both terms as equally validated, disputed, and refined could lead to distrust of *all* science. The literature suggests that there are three main factors at play that determine the use of scientific concepts in making decisions in everyday science: authenticity (found in perceived action), deep understanding of scientific concept, and the ability to connect the scientific concept with the decision issue. Furthermore, the research that has been conducted with science teachers (with science degrees) and scientists indicate they too gave priority to social concerns and personal values in everyday decision making (Aikenhead 2006, p. 104).

The differences in undergraduate student ability to perceive the role of evidence is substantiated in the literature of personal epistemology. William Perry (1970) and his successors have demonstrated intellectual and ethical developmental schemas (positions) that students go through during their four-year undergraduate experience. The following table shows the connection between Perry Position and perception of evidence as outlined in the reflective judgment model (King and Kitchener 2002).

Table 12.1. Perry Positions (Perry 1970) and Perception of Evidence (King and Kitchener 2002)

Perry Position	Perception of Evidence
Dualism (Positions 1–2): Right/wrong, good/bad, everything is known, Authorities know the "truth," and knowledge is "right" answers.	No need for justification or evidence. Knowledge is certain and right. Authority is the source of knowledge.
Multiplicity (Positions 3–4): Students begin to see that there is room for human uncertainty, but there still is Truth. (4a) All opinions equally valid if Authority doesn't have the answer. Or (4b) recognizes ambiguity but also begins to incorporate ideas of evidence and context. Goes from what Authority wants to how Authority wants me to think.	Idiosyncratic beliefs, rules of inquiry—it is all based on social norms.
Contextual Relativism (Positions 5–6): Knowledge as contextual and relativistic. "Authority" is open to debate, analysis, and evaluation. Metacognition is developed here.	Evaluation of objective evidence through personal criteria (comparing evidence and opinion); knowledge requires action of the learner and is based on information from a variety of sources.
Commitment to Relativism (Dialectic): Developing personal commitment is seen as a path to identity.	Evaluation via empirical data and rules of inquiry—logic; subject is involved in constructing knowledge and is aware that knowledge is tentative; beliefs are justified on the basis of probability.

Researchers who use socioscientific issues (SSI) to explore student decision making. These issues are informed by scientific evidence but include social factors (political, religious, aesthetic, economic, ecological, ethical, and so on). "SSI represents ill-structured problems that lack clear-cut solutions. They deal with challenging premises, unknown parameters, and ethical quandaries" (Topcu, Sadler, and Yilmaz-Tuzan 2010). The use of SSI in instruction increases student engagement through authentic, relevant options and gives the opportunity to develop scientifically literate citizens who can "identify scientific issues underlying national and local decisions and express positions that are scientifically and technologically informed" (NSTA 2000). Wilmes and Howarth (2009) characterized the issue-oriented science classroom versus more traditional methods in Table 12.2 (p. 162).

Table 12.2. Characteristics Involved in Issue-Oriented Science Classrooms (Wilmes and Howarth 2009)

Less Emphasis on ...	More Emphasis on ...
Discussing science in isolation	Discussing science concepts and understanding in the context of personal and societal issues
Working alone	Collaborating with a group that simulates the work of a scientific community or represents authentic groups found in society
Acquiring scientific information	Acquiring conceptual understanding and applying information and conceptual understanding in making and evaluating personal, societal, and global decisions.
Closed questions with one correct answer	Open-ended questions that require students to explain phenomena or take positions backed by evidence
Multiple-choice assessments	Authentic assessments

Toulmin's Argument Pattern (TAP) is an argumentation evaluation, analysis, and practice tool used in science education as well as across a range of disciplines (Topcu, Sadler, and Yilmaz-Tuzan 2010). Erduran, Simon, and Osborne (2004) describe the structural framework of TAP as "an interconnected set of a claim; data that support that claim; warrants that provide a link between the data and the claim; backings that strengthen the warrants; and finally, rebuttals which point to the circumstances under which the claim would not hold true" (p. 917). The authors point out that while TAP defines argument, there is difficulty in using the framework because of the ambiguity involved in what counts as each component (claim, data, warrant) during actual use. They used TAP to code discourse as part of an investigation of classroom (or small group) argumentation improvement. For the purposes of introducing argumentation through debate, I have found the TAP helpful. Recent literature on debate in the science classroom is limited, and assessment often involves student self-report. Scott (2009) suggests that while critical thinking is a stated goal in many university courses, time and resources to develop, enact, and reflect upon instruction are minimal. Instruction in critical thinking and argumentation, as in other areas, requires suitable attention to planning in order to achieve positive outcomes and must go beyond traditional modes of passive learning strategies used in higher education. Students must be prepared and explicitly involved in the debate process at every step (Proulx 2004). Scott (2009) developed debate modules based on Bloom's Taxonomy and prepared students by introducing the debate process and an assignment sheet that included instructions on preparing a case brief. Based on postquestionnaire responses, students believed that the debates helped them learn new knowledge, have a better understanding of the topic, increased critical thinking, and was a good experience. No additional assessment information was provided that would have substantiated the students' beliefs. Proulx (2004) prepared students by introducing them to scientific method and critical thinking, with explicit attention to identifying the main idea of the argument and evaluating the sources of information, evidence, and the claim itself. Instead of an actual debate, students evaluated a speaker's presentation using the criteria that had been introduced. The students' class evaluations suggested that they felt the integration of critical thinking into the course and the presentation were valuable learning experiences and opened their minds to evaluation of differing views. The author stated that he felt that the addition of scientific method and critical thinking into the debate of environmental issues was "advantageous" since students were able to give justification for and articulate their point of view. Proulx also mentioned the importance of and difficulty in getting students that are emotionally engaged with the topic to listen, focus, and understand opposing viewpoints.

As mentioned previously, the great debate came out of frustration with how few outcomes were being realized with an instructional strategy (debate) that is being used in classrooms at all levels but seldom with intention and rigor. In order to get maximum results, instructors must use assessment to inform their practice, know what they are assessing (objectives), and intentionally plan for outcomes. Debate can't deliver critical thinkers without instruction in critical thinking.

The great debate has been developed over several years with variation depending upon context and student interest. I have used it in many different biology and education courses as well as the environmental education program leading to specialized certification that is used as the context here.

Initially, the great debate was used in the environmental education teaching methods course as an example of an activity that could be easily revised for more learning impact. However, after noticing discrepancies in student preparation with the actual activity objectives (i.e., development of critical-thinking skills, content, and argumentation), the activity was moved to the prerequisite course, introduction to environmental education, where the focus was on individual development of these skills instead of the pedagogy. The activity could then be revisited in the methods course from a pedagogical perspective and possibly, if they chose, included in what they actually taught during their capstone experience.

Major Facets

Day 1

Prior to announcing the debate, we explore our underlying assumptions and values that are major determinants in developing opinions on socioscientific issues through a modified version of a Project Learning Tree (1997) activity called Values on the Line. Students are given a copy of 10 statements made about local environmental issues and asked to rank to what extent they agree or disagree with each statement on a scale from 1 to 10 (10 being highest agreement and 1 being highest disagreement). Students are next asked to reread statement #1 and position themselves according to their ranking on a number line (from 1 to 10) that has been placed on the floor. The midpoint in the range of opinions is found and one side stays fixed while the other side moves across to stand in front of those people that remained in place. Each person has a partner who has ranked the statement differently. (In some cases, this may not happen. However, I have found that even partners that ranked the statement the same have different reasons for doing so.) One line of students is given 45 seconds to state their level of agreement or disagreement with the statement and then give justification for why they chose that level of agreement. Next, their partner in the other line is given 15 seconds to paraphrase and ask for clarification, if needed, of what was said and another 15 seconds is given to the first student to make that clarification and correct any misinterpretation of the original statement. At this point the partners switch roles, giving the other person a chance to explain their agreement or disagreement and for paraphrasing and clarification to take place. Students are asked:

- What reasons did you or your partners give for their rankings?
- Did any of the reasons make you want to change your mind?
- What were some of the stronger reasons? What were some of the weaker reasons?
- What made the reasons stronger and/or weaker?
- Were there recognizable underlying assumptions/values implied in the statements? Where did these come from? What kinds of experiences change or strengthen people's values?

The procedure above is repeated for three or four more statements. At this time students are asked to come up with at least three reasons to justify their chosen position on the next statement. After both people have made their statement and justifications and clarification has taken place, one minute is given to think over/write down their partner's three justifications and come up with rebuttals for each. Each person gets 45 seconds to provide the rebuttals to their partners stated justifications, with an additional round to address the rebuttals (30 seconds). Students are asked:

- How did it work with the additional component?
- What rebuttals were given? What were some strong rebuttals? What were some weak rebuttals?
- How could these arguments be strengthened?

Question prompts are used in a way to reflect the 5E Learning Cycle *explore* and *explain* stages. This involves asking students what they learned as they were exploring (the interaction in this case) and leading them to explain the concepts of the "lesson." The instructor's job is to fill in the gaps of or make explicit student knowledge as it relates to the objectives. Here, that involves students being able to:

- describe how attitudes and viewpoints are influenced by value factors (political, economic, religious, social, aesthetic, ethical, and so on) learned from past experiences
- identify the parts of an argument
- analyze the parts of an argument for consistency, strengths, and weaknesses
- anticipate counter argument(s) and provide quality rebuttal(s)
- evaluate sources of information
- identify, research, evaluate, and utilize the scientific components of a socioscientific argument

Students are engaged in discussion and given information on seeking out credible sources (primary and secondary articles; web-based searches), common fallacies of logic, factors that influence decision making, and a simplistic argument talk frame (Anonymous, n.d.) to help guide their initial research and preparation for argumentation. Even though some would consider this talk frame too simplistic to use in higher education, I have found that it is a good concrete guide for those who are unfamiliar with argumentation and it results in better initial preparation. The talk frame is vertical, with five "frames," with the following statements in each frame: My idea is that... (Position); My reasons are that... (Rationale); Arguments against my idea might be that... (Counter-Position); I would convince somebody that does not believe me by... (Rebuttal); and, The evidence I would use to convince them is that... (Rebuttal Rationale). Enough room is given in each frame for students to write in their initial responses.

The announcement is made that we are going to have the "great debate" and a count-off takes place with "1s" being pro and "2s" being against the statement. The statement can be derived from student discussion, the Values on the Line statements that were not used, or another place, but it should be relevant to the students, connect to current course content, and allow for multiple perspectives. In addition, student groupings need to reflect the size of the class and the intent of the debate. My focus for the great debate is on the process, not on who wins or loses the actual debate. In very large classes, I might have everyone prepare and choose 6–8 students to actually perform the debate for the rest of the class. Or I might have two or three debate groups working on separate statements. Regardless, efficiency and effectiveness are the most important criteria when choosing how to set up the great debate.

Students are instructed to research their position, complete two copies of their talking frame (one to turn in and one to work from), and come prepared with three rationale/arguments (one science-based), one counter position, a rebuttal, and a rebuttal rationale for next time. Remind the students that as they gather facts, they need to use credible sources and keep a reference list. They should keep notes as to how credible the source is. It is important to let the students know that points will be given for the quality and completion of the talking frame.

Day 2

Collect copies of the talking frame at the beginning of class. Have students meet with their groups. Instruct students to quickly share their individual work and compile a master list and rank the arguments from strongest to weakest. Support the students (as needed) in developing and using criteria. Students generally have difficulty with the idea that some evidence is better than other evidence. This point needs to be discussed explicitly. Have students evaluate the possible counter-positions and their rebuttals. Do they have the information necessary to address the counter-position? Have each group determine what research needs to be complete by the next class meeting. This should take no longer than 10 minutes of class time. Have students write their thoughts about the meeting and project in their learning logs (see assessment).

Day 3

Tell the students they will have 15 minutes to complete today's task and prepare for the debate for next time. Stating time limits and sticking to them give a sense of urgency to the task and keeps students focused. I also suggest that students take notes today. Ask the students to get into their pro and con groups and once there count off by two. The "1s" stay in their group and the "2s" switch sides. Students by now have developed a community with their original group and may voice their disapproval. I have found that this is necessary to keep the focus on the argument and not on winning or losing. It is important to explain to them that in a court of law the prosecuting and defending attorneys are required to share evidence that they will submit along with a witness list. This enables both sides to examine the evidence carefully to weigh how credible it is and to look for possible holes in the argument. Here you expect them to do the same. Now that both groups have individuals from both former sides for and against the statement, they should reexamine each argument component. Is there new information that the people who were formally on the other side can share about the opposite side's argument? Were some of the counter-points from the other side part of this group's original arguments? Are the sources of information the same or different? Was there agreement in the credibility of sources? Does the new information help your new side make a stronger case? How?

Once the class is brought back together, ask them what they learned from their work on the project. (Use the objective to frame prompting questions as necessary.) Give them the structure for the debate (and the rubric) and inform them that they should be prepared to argue either side. An extra individual point will be given for correctly identifying fallacies of logic used during the debate. In addition, a short written assignment (due following the class after the debate) is given in which the students address the following questions:

- Did you initially agree or disagree with the debate statement? What factors and values was this based on? Please explain. Did you change your position during or following the debate project? Why or why not? Please explain.
- What three arguments in the debate did you feel were the strongest? Why were they strong? What were their weaknesses? (What rebuttal might you have offered to them if given during the debate?)
- Evaluate the quality of the actual debate. Provide the criteria that you used along with at least two concrete examples.
- What did you learn about [the science concept] as a result of this project? What in the project helped with learning?

Day 4

Have students count off by 2s to determine position, pro or con. Give sides five minutes to choose their speakers and plan their opening statement. Remind students of the debate procedures. Regardless of which debate style that is chosen, it is important to offer work time between each component for the individuals to consult and plan their rebuttals and responses. I have found that this improves the thinking and allows more voices in the decision-making process.

Extensions

The first debate is always the most time-consuming because of the amount of new information and organization required. It is also where the most learning occurs. I highly recommend at least one more opportunity to practice their new skills and the opportunity to further reflect and synthesize ideas of critical thinking and argumentation.

Evidence for Success

Both formative and summative evaluation is used throughout the great debate project. Peer review is also used as time permits. Methods of formative evaluation include questioning, learning log entries, exit slips, and the argument talking frame. Formative evaluation gives both the student and instructor the opportunity to see where students are and how to improve to meet the objectives as well as informing subsequent instruction.

The questioning framework ties to the objectives of the activity and in addition to evaluation, is used to drive instruction. It can be used as a model of critical thinking and to develop critical-thinking skills if used explicitly with students. To do this, I ask students to listen to the types of questions I am asking to improve the argument and encourage them to ask those types of questions themselves as they are working. Students do not automatically show their thinking and so the instructor needs to ask them to do this explicitly, either through writing assignments (learning logs, exit slips) or verbally in class. Improvement in developing metacognition (thinking about their thinking) is dependent upon the extent to which the instructor focuses on it as an outcome during the project, the amount of time devoted to the project and/or supporting activities, and the student's level of development. Students in lower levels of development (dualism and multiplicity) need to be well supported with questions such as, "Where did you get that idea?" "What are the strengths and weaknesses in that idea?" or "What questions do we need to (or what questions will the other side) ask about that idea?"

These students are motivated by "winning" the debate, points (although they generally don't understand how many points will influence a final grade), and peers. They don't see a need for learning how to think because they see knowledge as coming from Authority and being right (or if it isn't right or correct then all opinions are equal). These people become frustrated with the ill-structured and ambiguous framework of the project. They do respond when given concrete directions, modeled questions to ask themselves, and opportunities for peer interaction (especially with students at higher developmental levels). Evidence of success through questioning is in the evolution of answers. In general, at the beginning of the project, those in lower developmental levels are unable to articulate their ideas and often resort to "I don't know" statements. (This could be a learned response since instructors usually move on to someone else who can answer the question indicating that thinking things out is not important.) From here, students start at different places but move along a continuum from giving their opinion only (with no reasoning even when asked); giving opinions with a reason based on personal experience or anecdotal information; giving opinions based on other sources; giving opinions with reasons based on credible sources and being able to point out the weaknesses (at this stage students are also able to point out weaknesses in other arguments using the same tools); and, last, to give a justified opinion, counterpoints and rebuttals based on credible evidence, and to rank evidence according to strength.

There are noticeable changes in thinking as the project continues, with the greatest changes in those in lower developmental levels. There is also improvement in articulation in those at higher developmental levels as a result of what could be the demand for communicating with and preparing team members for the rigor of the debate. For this reason I have found it necessary not to divulge the exact protocol for the debate at the beginning of the project. Students in higher developmental levels will not expend energy on people that may be on the opposite team even though their own learning is often enhanced through explanation to others.

Summative evaluation included the short paper that is turned in following the debate and the class evaluation. In comparing the individual student work from the argument talking frame and the short assignment following the debate (in a recent course), the improvement is noticeable for all 11 students. Four of the students completed the talking frame with good quality answers. Only two of these four attempted to use science understandings in the argument but all four did provide information from sources (as opposed to personal experience). Five students answered all five frames but had difficulty addressing how they "would convince somebody that does not believe" them and offering evidence to convince them. Three of these people did not address the argument against their idea when trying to convince the person that didn't believe them, showing a lack of understanding the importance of consistency in an argument. The evidence used for four of the five was weak and based on personal experience and values. Two students articulated their ideas about the statement but offered reasons based on personal value systems and a weak counter-position statement. In the final paper, all of the 11 participants were able to articulate their position on the statement and give a quality rationale for their position. Five people changed their position and three gave a thoughtful rationale as to how their thinking had changed (showing metacognition). Two individuals who changed their position offered that they did so only because they were convinced by their teammates that it was the "right" position (these people were on opposite sides). There was no indication at what time they switched their position and their answers suggested that they were mimicking the statements and rationales of the team. This, to me, indicates success in that these people are being pushed by the activity to see complexity and decision-making patterns even if it hasn't been assimilated into their way of thinking. Seven of the participants were able to identify, rank and evaluate the strength and weaknesses of the arguments and included the opposing team's argument. The other four chose arguments from their own team to identify and rank but did not conduct a quality evaluation. All 11 said they learned more about the science topic but only six gave what was specifically learned and what in the project helped them.

I have also used another assignment that involves writing opinion pieces on three science research articles at the beginning, middle, and end of the course to look for improvement in critical thinking and argumentation. While the improvement is noticeable for the majority of students, it would be difficult to attribute it solely to the great debate activity. As with all assignments, students get much feedback and are allowed to revise and resubmit. The great debate itself is assessed and could be included as a summative assessment. However, the rubric is designed to give group points and individual improvement in thinking is not measured. When time permits having more than one debate during the semester, there is a noticeable improvement in speaking and the argument. In addition, anecdotally the results of final exams seem to suggest that students, in general, performed better on those topics that had connected debates. Statistical analysis has not been performed.

The learning logs and the course evaluations indicate that the students feel that the great debate is a good learning opportunity and something they enjoyed doing. Some comments were:

- "I loved the debates. We never get a chance to discuss our ideas in other classes and it was interesting to see what the other students thought."

- "I didn't like the debate at first because it was a lot of work. Now I like it because I can tell my friends when they aren't making sense. Just kidding. I learned a lot."
- "Telling others how to do their work was frustrating but it helped me realize what I needed to learn more about. It turned out to be a good project. There should have been more points for it and those that did more of the work."
- "The debate and getting ready for the debate was my favorite part of the class. I wish we could have had more time to do more of them."

Next Steps

As the internet and social media have become more prevalent in our society, higher education, especially in undergraduate education, has not kept up with the meeting the education demands that our students will face in the 21st century. It is vital that we teach and incorporate into our lessons how to evaluate the overwhelming amount of information coming at us and all of the forms it is in. In the future I plan to augment instruction with web-based applications. There are valid instruments that have been developed in the last few years that measure internet literacy skills and many of the examples used are from science websites.

The evaluation of this type of project has many limitations, the most important being the lack of opportunity for controls, the ability to limit variables in a real classroom situation, and the time to administer pre/post-instruments to measure learning outside actual science content. Keeping in mind the commitment that science educators and scientists must make of increasing scientific literacy among all citizens in our democracy, it is absolutely necessary to find ways to efficiently and effectively incorporate science literacy and with it critical-thinking skills and nature of science understanding. The information collected from this project has suggested that there are improved outcomes in the development of critical thinking and science literacy. However, quasi-experimental research studies would be necessary to determine the factors involved and the extent to which they could influence outcomes.

Specific Ties to Other Specific Reform Efforts

In addition to the NSES (1996), this project meets the 21st Century Skills for Science (Partnership for 21st Century Skills 2009) and ties into the *Vision and Change in Undergraduate Biology Education* document (Brewer and Smith 2009). The 21st Century Skills for Science calls for creativity and innovation (including "application of theory to real world situations" and investigations that are more interdisciplinary); critical thinking and problem solving; communication; collaboration; information literacy; media literacy (interpretation for both the scientific community and the general public); information and communications technology literacy; flexibility and adaptability (directly ties into nature of science); initiative and self-direction; social and cross-cultural skills; productivity and accountability; and leadership and responsibility.

At the core of *Vision and Change in Undergraduate Biology Education* is the necessity for student-centered classrooms and learning outcomes that are "interactive, inquiry driven, cooperative, collaborative, and relevant." Students should receive continual feedback, be encouraged and empowered to strengthen their thinking and be given multiple opportunities to evaluate complex problems from a variety of perspectives.

With several iterations of the great debate, I have tried to meet these program and learning goals. This strategy has been successful in providing students with opportunities to investigate relevant and authentic local problems, develop critical-thinking skills, and take responsibility for their own learning and how they demonstrate it. With more refinement and investigation into the individual components of the great debate, it will continue to provide greater learning outcomes with more efficiency.

References

Aikenhead, G. S. 2006. *Science education for everyday life: Evidence-based practice.* New York: Teachers College Press.

Anonymous. n.d. An argument "talk frame." *www.azteachscience.co.uk/ext/cpd/dips/resources/pdfs/Talk-frame.pdf*

Brewer, C. A., and D. Smith, ed. 2009. *Vision and change in undergraduate biology education: A call for action.* Washington, DC: National Academies Press.

Erduran, S., D. Ardac, and B. Yakmaci-Guzel. 2006. Learning to teach argumentation: Case studies of preservice secondary science teachers. *Eurasia Journal of Mathematics, Science, and Technology Education* 2 (2): 1–14.

Erduran, S., S. Simon, and J. Osborne. 2004. TAPping into argumentation: Developments in Toulmin's argument pattern for studying science discourse. *Science Education* 88: 915–933.

Kelly, G. J. 2007. Discourse in science classrooms. In *Handbook of research on science education*, ed. S. K. Abell and N. G. Lederman, 443–469. Mahwah, NJ: Lawrence Erlbaum.

King, P. M., and K. S. Kitchener. 2002. The reflective judgment model: Twenty years of research on epistemic cognition. In *Personal epistemology: The psychology of beliefs about knowledge and knowing*, ed. B. K. Hofer, and P. R. Pintrich, 37–61. Mahway, NJ: Lawrence Erlbaum.

National Research Council (NRC). 1996. *National science education standards*. Washington, DC: National Academies Press.

National Science Teachers Association (NSTA). 2000. Beyond 2000—Teachers of science speak out. *www.nsta.org/about/positions/beyond2000.aspx*

Osborne, J. F. 2010. An argument for arguments in science class. *Kappan* 91: 62–65.

Partnership for 21st Century Skills. 2009. 21st Century skills map—Science. *www.p21.org/storage/documents/21stcskillsmap_science.pdf*

Perry, W. G. Jr. 1970. *Forms of intellectual and ethical development in the college years: A scheme*. Troy, MO: Holt, Rinehart and Winston.

Proulx, G. 2004. Integrating scientific method and critical thinking in classroom debates on environmental issues. *The American Biology Teacher* 66: 26–33.

Sampson, V., and J. Grooms. 2009. Promoting and supporting scientific argumentation in the classroom: The evaluate-alternatives instructional model. *Science Scope* 33: 67–73.

Sampson, V., and J. Grooms. 2010. Generate an argument: An instructional model. *The Science Teacher* 77 (5): 33–37.

Scott, S. 2009. Perceptions of students' learning critical thinking through debate in a technology classroom: A case study. *The Journal of Technology Studies* 34: 39–44.

Shulman, L. 1987. Knowledge and teaching: Foundations of a new reform. *Harvard Educational Review* 57: 1–22.

Topcu, M. S., T. D. Sadler, and O. Yilmaz-Tuzan. 2010. Preservice science teachers' informal reasoning about socioscientific issues: The influence of issue context. *International Journal of Science Education* 32: 2475–2495.

Wilmes, S., and J. Howarth. 2009. Using issue-based science in the classroom. *The Science Teacher* 76 (7): 24–29.

The Student-Centered Lecture: Incorporating Inquiry in Large-Group Settings

Holly J. Travis
Indiana University of Pennsylvania

Setting

Indiana University of Pennsylvania (IUP) is the largest of the Pennsylvania State System of Higher Education (PASSHE) schools. Located in a rural area of western Pennsylvania an hour north of Pittsburgh, it serves students from all over the state, as well as many out-of-state and international students. Originally founded as a teacher's college, IUP has expanded its offerings to include more than 100 majors in areas such as fine arts, humanities, social sciences, nursing, and natural sciences, as well as a broad range of education programs, at both undergraduate and graduate levels.

With an enrollment of 15,000 students, it is very common to have 48, 72, 120 or more students in a single lecture section of the liberal studies science courses such as general biology I and general biology II. The same can be true for service courses offered to specific majors, such as the science courses designed to meet the requirements of the elementary education curriculum, including fundamentals of environmental biology, which serves as the life science content course for these students. Other classes, such as basic biology—which serves as a preparatory class for students who enter the university as biology majors but demonstrate through placement exams and SAT scores that they may be lacking some critical areas of preparation—tend to have much smaller enrollment, generally numbering 24 students or less. Students in all of these varied classes, however, need to be prepared to think critically and evaluate information effectively.

Overview

Every science educator has heard about the importance of inquiry not only for effective science teaching but also for meeting state and national standards. The National Science Education Standards (NSES) note that "students develop an understanding of the natural world when they are actively engaged in scientific inquiry" (NRC 1996, p. 29). The NSES also stress the importance of instructors as guides who should be facilitating student learning, rather than simply spouting information while students sit passively in the classroom. Active learning means that the students should be interacting with other students and the instructor, that they are challenging their own preconceived ideas, and that they are taking some level of responsibility for their own learning (National Research Council, 1996). This is also stressed in *Vision and Change in Undergraduate Biology Education: A Call for Action* (Brewer and Smith 2009), a document promoted by the American Association for the Advancement of Science (AAAS), the National Science Foundation (NSF) and the National Research Council (NRC). These reports specifically note that undergraduate students should engage in inquiry-based learning that relates to the real world and focuses on conceptual understanding rather than on the memorization of facts (Brewer and Smith 2009). Many tools, such as the BSCS 5E Instructional Model, have been created to improve teaching and learning in elementary and secondary school classrooms, but the traditional large lecture classes so often observed in higher education settings can make this very difficult to manage.

At IUP, instructors of various science classes offered in the biology department have made an effort to incorporate inquiry and critical thinking into their classes. These include majors courses, nonmajors liberal studies science offerings, and service courses designed to meet the needs of specific programs. Several different strategies have been tried in laboratory courses, small seminar or classroom sections, and large lecture settings, some more successful than others. For one instructor, an example that has proven to be valuable in many different classroom settings and for many different student audiences is a modification of the 5E model. This strategy has allowed students to benefit from active involvement in discussions while accessing prior knowledge and working through their misconceptions as they incorporate new information introduced in class.

The smallest class to use this modified version of the 5E model was basic biology. As noted above, this course is designed to assist underprepared students who have enrolled at IUP as biology majors in a variety of programs, including premed, prevet, and biology education, to strengthen the skills and background knowledge necessary for success in these demanding majors programs. This class generally has an enrollment of 15–24 students, and is required for students identified as underprepared, based on SAT scores and results on the mathematics placement exam taken by all incoming students. These students must pass basic biology with a grade of C or better before they can move on to the principles of biology I and II courses.

Another group of students who have been exposed to this teaching strategy are the elementary education majors taking their life science content course. All students in this program are required to take a specially designed series of science courses in physics, chemistry, Earth science, and life science in order to be prepared for their certification exams and for their future role as classroom teachers. Their life science course, fundamentals of environmental biology for elementary education majors, is a service course offered specifically for these students, and it generally has an enrollment of 35–65 participants. There are dual benefits for this group of students, as they are not only active participants in critical thinking and active learning, but they also have the opportunity to observe student-centered learning in action. This will make it easier for them to use these techniques in their own classrooms, since they have seen it modeled and have been actively involved during the semester.

The final and largest university courses to take advantage of this teaching strategy are general biology I and general biology II sections offered at both the main campus and at the Northpointe branch campus. These are liberal studies of science offerings that cover a wide range of topics. General biology I, offered during the fall semester, includes ecology, basic biochemistry, cells, energy, genetics, and evolution. General biology II, offered during the spring semester, covers plant and animal body systems, with a primary focus on human biology. Course enrollments in the sections utilizing the student-centered teaching strategies discussed here generally range from 48–120 students.

An Overview of the 5E Model

The BSCS 5E instructional scheme was developed by Trowbridge and Bybee (1990) as a means of easily creating a student-centered lesson for elementary and secondary classrooms. It is made up of five basic sections: *engage, explore, explain, elaborate, and evaluate.* It should be mentioned that these sections can be repeated within a lesson as needed to cover the required material.

In a traditional K–12 classroom setting, each new topic is introduced with an appropriate *engage*. This is a brief activity, question, demonstration, or film "snippet" that whets the students' appetite and gets their attention focused on the topic. A new engage activity should be done for each different topic that is introduced within a lesson, or when a topic must be continued during another class period. Once the students have seen or done the engage, they are asked to *explore* the topic in groups. This might be a lab activity in which the students develop questions and attempt to answer them, or it may include several challenge questions that encourage the students to think in depth about the topic at hand. After students have had time to explore the topic and to develop answers, they are asked to *explain* these answers to the whole class. Teams take turns sharing their results, whether they have done an experiment or worked through a question. Different team members are encouraged to present the group's material so all students participate in the discussion. At this point, the instructor can also address misconceptions and misinformation to ensure that students understand the material. Once the groups have had a chance to explain the information in their own words and have a clear understanding of the material, they *elaborate* on the topic. They may do further laboratory work, do research, give presentations, or simply discuss more complex questions within their groups. This allows them to build a deeper understanding and to relate this information to other material covered in class. Finally, the instructor (and students!) must *evaluate* student comprehension of the topic. Evaluation can take numerous forms, including standard quizzes or tests, written assignments, oral presentations, student self-evaluations or observation of student participation in group activities.

The 5E model aids the instructor in maintaining a smooth flow during a class by giving a simple outline for developing the class procedures. It reduces time spent on unrelated topics and helps keep student groups on task. It also aids in developing better understanding of the material by encouraging students to explore new information before the explanation and by encouraging a variety of assessment techniques. Instructors guide, focus, challenge, and encourage student learning as they move through these steps (Wilder and Shuttleworth 2005).

Translating the 5E Model Into a Higher Education Setting

In higher-education settings, especially in science courses, classes have traditionally been offered as didactic lectures. Students are expected to sit quietly, paying attention to the instructor and taking diligent notes on the information covered, whether it is spoken or placed on a screen in the front of the classroom in the form of overhead transparencies, or more recently, PowerPoint presentations. Questions may or may not be encouraged, depending on the instructor, and discussion with other students in the class is unmistakably discouraged, at least during the lecture. Classroom design also inhibits interaction, with fixed seating and a single large screen placed at the front of the room. This passive lecture style encourages students to simply memorize facts and formulas, leading to decreased retention and poor understanding of major principles and concepts (Baldwin 2009; Wood 2009). By making some changes in the way the classroom setting is used, a traditional college lecture can become a dynamic setting for student learning. This focus on a student-centered classroom is specifically recommended in the *Vision and Change* document, which stresses classroom practices such as student participation, multiple forms of assessment, and using cooperative learning strategies (Brewer and Smith 2009). Adding discussion and student interactions to the lecture setting creates the development of a more energetic atmosphere. Structuring information in a form that encourages inquiry and critical thinking helps students successfully integrate new information into their existing background knowledge. This leads to increased understandings and retention of the information, whether students are biology majors headed to upper-level science classes, future educators who need a solid grasp of content in order to effectively teach in an elementary classroom, or nonmajors who simply need to be informed citizens capable of understanding and evaluating issues facing them in the future.

General Changes to a Lecture Format

Several changes have been made to the general format of the large lecture to encourage thinking and discussion. One of the key changes is the promotion of student interaction throughout the class. This is not to say that students should be talking to each other all the time about unrelated topics and ignoring the instructor, but rather, that they interact frequently with others sitting nearby to discuss questions posed about content and prior knowledge. As noted in NSES, "learning science is something students do, not something that is done to them" (NRC 1996, p. 20). By allowing students to discuss their answers in pairs or small groups, students have the opportunity to access previous information they have learned and to start identifying misconceptions they may have. Encouraging peer interaction before asking for responses also helps students feel less self-conscious about potentially being incorrect, since they have had additional input and do not feel as if they are the only one who doesn't know the proper answer.

A second difference between the traditional lecture and the student-centered classroom is the emphasis on having the "right" answer. In a traditional lecture class, students who volunteer answers to questions posed by the instructor either have the correct response or the wrong one. Generally, a very specific response is sought by the instructor and students who may not be sure are discouraged about offering their thoughts for fear of being embarrassed if they are wrong. In a student-centered class, students' answers, whether correct or incorrect, are used to uncover misconceptions and to provide a framework that is then used by the instructor to reconstruct knowledge around the appropriate information. Incorrect responses may actually be more valuable for class discussion, as they can lead to additional questions that assist the students in recognizing inconsistencies between what they already know and what is being introduced in class.

Finally, because students need some accountability and the instructor needs a way to assess student learning and evaluate misconceptions, an informal response paper is turned in by students each day. This is used for participation points and is not graded. It serves as a very valuable tool, giving the instructor the opportunity to see what background information students have retained from high school or previous science courses, and to note what material they still don't seem to be grasping effectively. The most difficult aspect of this daily paper is getting students to realize that incorrect answers are acceptable, and are, in fact, more useful than correct ones for evaluating key facets of the class! These papers are not returned to the students and should not be used for their notes. They are used solely by the instructor as a guide for understanding student learning and background knowledge.

These fairly simple changes to the typical lecture have a tremendous impact on the atmosphere of the classroom, and on the attentiveness of the students. Because they are constantly working with other students in the class and writing responses, students are less likely to doze off, play games on their phones or computers, or simply "zone out." They are also more comfortable asking questions because they are accustomed to interacting with the instructor throughout the class. Having to hand in the daily papers also ensures that they take time to think about the material being covered and write responses down. These changes, along with the use of steps from the 5E model, combine to make lectures interesting and student-centered.

Modifying the 5E Model for Large-Group Instruction

As discussed earlier in this chapter, there are five steps involved in the 5E model: engage, explore, explain, elaborate, and evaluate. Some of these may fit into most classes, whether they are didactic lectures or interactive seminars. For example, most classes start out with something designed to get the students' attention, whether it is a formal *engage*, such as a movie clip or opening question, a reminder about assignments due in the near future, or simply a greeting from the instructor. In a traditional college lecture, this might take the form of a statement reminding students where they left off in the previous class or an introduction to a new topic. In the modified lecture, this is generally either a review question to remind students what material was being covered during the previous class, or a question designed to have them start thinking about new material (see Figure 13.1, p. 176).

One of the most important features of the 5E model is the *exploration* of new material before it is *explained*. This key change in traditional lecture format opens the door to an inquiry-based class and involves students in their own learning. For example, in starting a chapter on the properties of water, the first turn-in paper question given to students might be to "List 10 things you know about water." They can easily come up with two or three items, but then they have to start thinking a little harder and discussing options with other students sitting nearby. They generally come up with the inevitable "it's wet" statement, thinking that they are being funny.

Figure 13.1. Engage

The Instructor poses a question to the entire group. Students work together in small groups with others seated nearby to come up with responses.

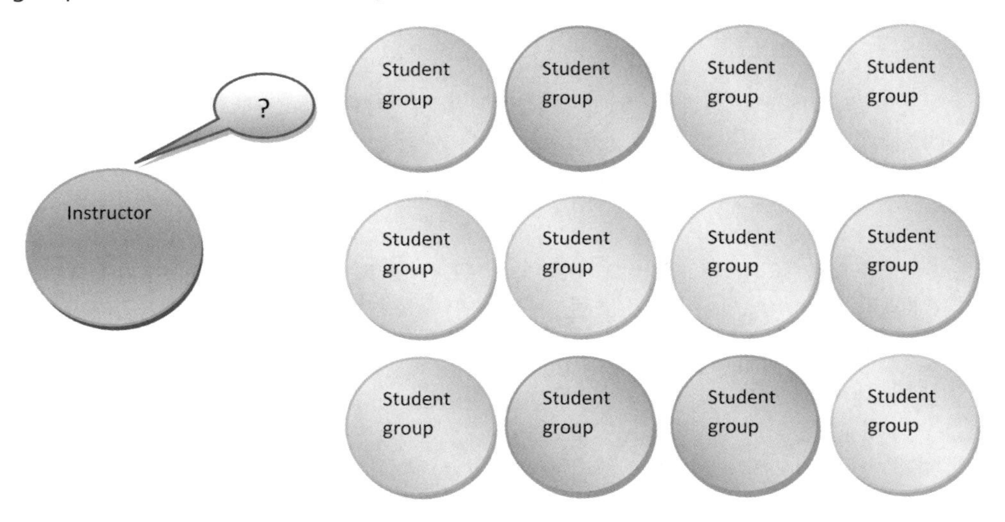

This can, however, lead to a discussion about water being one of a very few substances that is liquid at room temperature, and the importance of liquid water to living things, which is a very valid and important aspect of the analyses here. When a student mentions swimming, discussion of a belly flop and its relationship to surface tension can be introduced. Rather than simply listening to the instructor list the properties of water listed in the textbook, the students have provided valuable information and are much more interested because they feel that they contributed something to the class discussion. Putting the exploratory question first and basing the explanation of key terms and content on student contributions turns a dry science lecture into a dynamic student-centered discussion (see Figure 13.2).

Another important illustration of this format involves the interpretation of information provided in graphical form. Many people are unable to look at a graph or chart and effectively comprehend and evaluate the information it provides. By incorporating these regularly throughout the class and having students answer questions such as, "What does this graph tell you about global water distribution?" students learn to determine what information is provided, and more importantly, what the graph does *not* say. Because graphs and charts are used frequently in all settings, from newspapers to medical results to nutrition information, every citizen should be able to review material presented in this format and make decisions based on their evaluation.

Using this *explore-explain* sequence throughout the class period keeps students involved and interested in the material. By asking questions designed to have them think about previous knowledge and integrate new information introduced through class discussion, and by using student responses to questions as the basis for explaining new information, what would otherwise be an exercise in boredom for both the instructor and the students becomes an interesting exercise

Figure 13.2. Explore-Explain

The student groups share their answers with the class. The instructor uses these responses to guide student learning, introducing terminology and content based on student background and experiences.

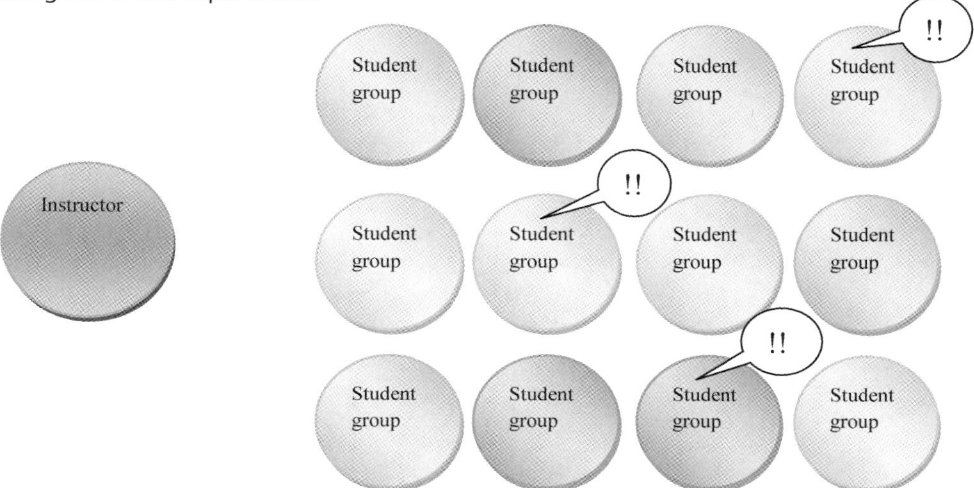

that relates science to the real world. This question-and-answer format, with the instructor posing questions to be answered on the turn-in papers and then using student responses as the basis for introduction of new content, is used throughout the class. Consequently, there are several explore-explain sequences in the course of a typical lecture class.

At the end of the class period or when the instructor has finished a key section of materials, a more advanced question that involves assimilating several terms or topics can be introduced. This would serve as the *elaborate* portion of the 5E model, asking students to build on previous knowledge and synthesize or evaluate related information. For example, after having students devise their own definitions of key terms such as *cohesion, adhesion*, and *surface tension* and discussing these as a class, the instructor might ask students to explain on their turn-in paper how water moves through a tree. After a discussion of diffusion, students might be asked the following: "A person becomes very dehydrated, so the concentration of water in her blood decreases. In which direction will water move across the plasma membranes of her blood cells? What will happen to the volume of the cells as a consequence? Why is proper hydration important for multicellular organisms?"

These types of questions require that students take the definitions and processes discussed and apply them to a real-life situation. Students can discuss this in pairs or small groups in order to help clarify their own thinking. Alternatively, these questions can be answered individually (without discussion in small groups) on the turn-in papers and can be used to *evaluate* students' understanding of the content. The correct response can then be discussed as the engage for the next lecture, giving the instructor the opportunity to clarify points that were unclear based on the student responses (see Figure 13.3, p. 178).

Figure 13.3. Elaborate and Evaluate

At the end of class, the instructor poses a more complex question, requiring that students assimilate information and terminology covered earlier. Responses can be shared with the entire group or submitted via individual turn-in papers to evaluate student understanding.

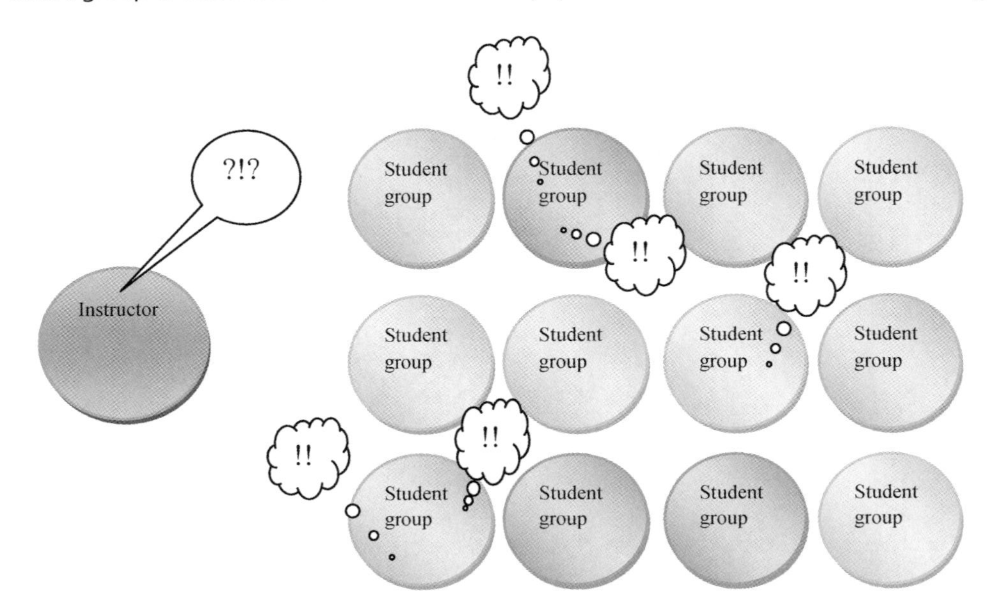

By using features of the 5E model throughout the class period, consistently engaging students in discussion with each other and with the instructor, and asking students to apply information covered in class to novel or real-life situations, content that would otherwise be considered boring and useless by typical students becomes interesting and relevant. This type of interaction occurs frequently in small upper-level seminar-type classes, but is typically ignored in large-group settings. Yet it is the students in these larger sections that benefit most from thinking about new information, collaborating with peers, and applying the content to real-life situations and questions. It typically engages students more, keeps them involved in the lecture, and gives the instructor frequent feedback on student understanding so material can be modified for more effective dissemination. It also improves student attitudes about the class, leading to lower levels of anxiety and more frequent student-instructor interaction.

Evidence Supporting Student-Centered Large-Group Instruction

Several items support this teaching format for use in lecture classes at all levels. Student evaluations, used by the department and the university for hiring, continuation, and promotion decisions, clearly listed the student-centered, interactive style of these classes as being a positive feature of all the courses taught by the instructor. Retention of students for additional courses and comments from students regarding their application of content in real-life situations outside the university, especially in the nonmajors' general biology class, also demonstrate its effectiveness.

Student Course Evaluations

Student evaluations, the bane of many a faculty member's existence, were one of the strongest indications that the interaction and active nature of these classes created a positive learning environment for the students. Comments from general biology students in response to the prompt, "List up to three things the instructor did that contributed to your learning" included statements such as

- "Used spontaneous questions to keep us paying attention (turn-in paper),"
- "Turn-in papers and questions that help clarify your class,"
- "The instructor helped clarify misunderstandings," or simply
- "Turn-in sheets," as noted on multiple response sheets.

In the class for elementary education majors, comments were often even more supportive. Examples of student comments included

- "Turn-in papers required us to think deeper into what we already know,"
- "The turn-in papers helped me evaluate what I didn't know and I was able to focus on those areas for the test,"
- "The turn-in papers are a great way to recall and think about information," and
- "Encourages students to explore what they know and how to apply it."

Students in the basic biology class said

- "The way she taught them was extremely useful and will definitely help me in my other courses," and
- "We had a turn in paper. It didn't sound very great at first but it is good review. We review things from other classes and turn in answers to questions that help us to piece information together. She has shown the best teaching method."

Overall instructor ratings for these courses were frequently ranked 90% and above and superior, and there were rarely any ratings that placed the instructor lower than average. Complaints tended to come from students who weren't using the turn-in papers as directed, probably because past experiences had taught them that only "correct" answers were acceptable. Examples include "turn-in papers make it difficult to take notes" and "sometimes I would be scrambling to put things on my turn in paper and then would not have time to write in my notes." Comments like these demonstrate that students are often afraid to turn in the wrong answer, even when they have been expressly told the reason behind this teaching strategy. Instead of writing what they already knew or what they had discussed with classmates on the turn-in paper, they still felt that they had to wait for the correct response, and then tried to hurriedly write it on both their turn-in paper and in their notes.

Student Retention in Additional Courses

In addition, especially at the branch campus, student retention in subsequent course sections taught by the same instructor was very high. General biology I and II, and general chemistry I are the typical liberal studies science classes offered at Northpointe during an academic year. Students have the option of taking two of these, or waiting until they move to main campus after one or two semesters and taking one or more of their science classes there, with course offerings available in biology, chemistry, geoscience, and physics. Nearly all students who took one of the general biology courses chose to take the other one as well, even when they had transferred to main campus and had to commute back to the branch campus for their second biology course. One student noted on their evaluation form that "I would probably take any class taught by [this instructor]."

Developing an Educated Community

Finally, student retention of the content covered in the courses remained high, as indicated by several students. This was especially noteworthy in the nonmajors' general biology courses, since the material did not necessarily apply to future coursework or career plans. One student made a point of contacting the instructor two years after taking general biology 104 to describe a situation where a family member was hospitalized for a heart problem. She commented that as the physician was explaining what happened, she realized that she understood what he was saying and could ask appropriate questions based on material covered in the biology class she had taken years before. Preparing a citizenry that is science-literate and able to think critically is arguably the most significant result of the interactive, student-centered lecture strategy, especially in the nonmajors' courses, where this may be their only exposure to the nature of science.

Future Directions for Teaching Science

As noted in the *Vision and Change* document, teaching methods in undergraduate science courses need to be updated in order to produce educated citizens prepared to take on the challenges they will face. While it requires an initial input of time and energy to update lecture notes and handouts and to integrate the use of new technologies, the time has come for instructors in higher education to take on these challenges. Not only do the students find these interactive teaching strategies more engaging, but they retain information better and learn valuable skills that can be applied to real-life situations they will encounter outside the classroom. This is especially important as science and technology continue to expand and develop at the remarkable pace seen today, because individuals must be able to evaluate information well past their time in the higher education setting, as individuals and as global citizens.

References

Baldwin, R. G. 2009. The climate for undergraduate teaching and learning in STEM fields. *New Directions for Teaching and Learning* 117: 9–17.

Brewer, C. A., and D. Smith, eds. 2009. *Vision and change in undergraduate biology education: A call for action*. Washington, DC: National Academies Press.

National Research Council (NRC). 1996. *National science education standards.* Washington, DC: National Academies Press.

Trowbridge, L., and R. Bybee. 1990. *Becoming a secondary school science teacher.* 5th ed. Englewood Cliffs, NJ: Merrill.

Wilder, M., and P. Shuttleworth. 2005. Cell inquiry: A 5E learning cycle lesson. *Science Activities* 41 (4): 36–43.

Wood, W. B. 2009. Innovations in teaching undergraduate biology and why we need them. *Annual Review of Cell and Developmental Biology* 25: 93–112.

Service Learning in an Undergraduate Introductory Environmental Science Course: Getting Students Involved With the Campus Community

Grace Eason
University of Maine, Farmington

Setting

As the public liberal arts college for the state of Maine, the University of Maine at Farmington (UMF) prepares students for engaged citizenship, enriching professional careers, and an enduring love of learning. Given its history since 1864 of educating teachers and its distinctive contemporary mission, UMF has consistently been rooted in a vigorous tradition of education in service of the public interest. In embracing this tradition, UMF seeks to graduate individuals who will live purposeful, ethical, and personally rewarding lives, and who will strengthen the social fabric of the communities they inhabit in Maine and beyond.

The university's focus is undergraduate education in a residential setting. An enrollment cap of approximately 2,000 students helps to support multiple modes of teaching and learning, but prioritizes face-to-face instruction with highly qualified faculty in settings that allow close relationships between students and their instructors to flourish. UMF also welcomes commuter students and provides limited graduate and continuing education opportunities where regional and statewide needs correspond with areas of academic strength in the university. Through its focus on high-quality academic programs in the arts and sciences, teacher education, and human services, the university challenges students to be active citizens in a campus community that helps them find and express with confidence their own voices, teaches them the humility to seek wisdom from others, and prepares them for ongoing explorations of how knowledge can be put to use for their personal benefit and the common good.

Overview

Why is it important for undergraduates to feel empowered in making a difference in their community? How can undergraduate institutions facilitate student involvement on campus, not just with extracurricular activities, but through the curriculum they are required to navigate in order to graduate? Can an introductory environmental science course for nonscience majors address these apparent needs? Consider the following quote:

> Beyond simply adding a few classes on environmental issues and sustainability, a growing cadre of individuals and organizations concerned about the fate of our ecological and social structures are calling for a fundamental rethinking of how institutions of higher education educate students, conduct research, interact with local communities and ecosystems, operate campuses, and provide a model for other institutions. Institutions of higher education clearly have the ability to be leaders in sustainable thought and practice (Shriberg 2004).

From global climate change to rising oil prices and the search for alternative energies, environmental education must be clearly modeled for students in order for them to incorporate what they have learned beyond their university experience and into their own lives. The National Science Education Standards (NSES) provide a framework for engaging students in environmental issues. For example, at the beginning of the Science in personal and social perspectives section is the following: "Central ideas related to health, populations, resources and environments provide the foundations for students' eventual understanding and actions as citizens" (NRC 1996, p. 138). There is a clear connection between the NSES and civic engagement. It is also emphasized throughout the standards that students do this through a steady diet of experiential learning of fewer, more in-depth concepts and appropriately progressing from concrete to formal concepts (Brown 2011).

In order to fulfill the goals stated within the NSES, the environmental education experience must be engaging as well as relevant for students, regardless of the type of degree they are pursuing. Dr. David Orr said it best:

> All education is environmental education. By what is included or excluded we teach students that they are part of or apart from the natural world. To teach economics, for example, without reference to the laws of thermodynamics or those of ecology is to teach a fundamentally important ecological lesson: that physics and ecology have nothing to do with the economy. That just happens to be dead wrong. The same is true throughout all of the curriculum. (Orr 1996, p. 52)

The challenge of course is to motivate and inspire students to realize that even though we face tremendous environmental challenges, actually doing something about these issues is possible even if it is on a smaller scale. Since UMF's undergraduate nonscience majors are required to take two natural science courses to fulfill their general education requirements, their experiences in this introductory environmental science course (ENV 110) might be the only time they have the opportunity to learn about these pressing environmental issues as well as act upon them.

Major Features

The main focus of this course is to guide students to reflect and consider whether or not it is possible to move beyond economic growth and pursue prosperity in their own communities (McKibben 2007). The population of students ($n=40$) taking this course consists of a variety of majors. Students learn not only about pressing environmental concerns and issues, but they are also provided with skills to become change agents within their communities. Service learning provides students with a place to start and a framework for them to make a difference in a small community setting. As described by Gutstein, Smith, and Manahan:

> Higher education pedagogy literature calls for a more engaged, participatory learning environment for undergraduates (Chickering and Gamson 1991) that includes student acquisition of transferable skills through some type of authentic experience (Chalkley and Harwood 1998). *The Dearing Report* (1997) reinforces this message, advocating that all undergraduate degree programs should include some element of vocational application. Service learning helps meet this need. (2006, p. 22)

It is challenging to motivate students who are required to take a science course to fulfill a general education requirement. It can also become increasingly challenging when students indicate that they are nervous or anxious about succeeding in a science course since it is not a primary focus in their major. One way to help students overcome their concerns regarding science is to gradually provide them with a framework on what science is before attempting to jump right into the content:

> The need for understanding of the nature of science for effective science teaching is generally accepted. There is also a need for an understanding of the relationship between science and environmental education, which draws on science to support knowledge of the causes of environmental problems, as well as the complexity of ecological systems. (Littledyke 1997, p. 643)

The severity and scope of environmental damage that students become aware of has the potential for putting students into a state of denial, anger and often times despair—feeling that nothing can really be done because all of these issues are just "too big" to deal with on an individual basis. This is where a small-scale service-learning project can help students realize that there is so much that can be done and all they have to do is shift from asking themselves "what's in it for me?" to asking instead, "how can I help?"

> Research into the effectiveness of environmental education has demonstrated that simply having knowledge of an issue doesn't result in behavioral change. Instead, for students to accept responsibility for the environment they need to take ownership of issues and feel empowered to do something about those issues. (Hungerford 1996, p. 27)

Service learning is a teaching strategy that provides students with that opportunity. It is an educational experience in which students participate in a service activity that meets a need within the community. Students gain further understanding of the course content, a broader appreciation of the discipline, and an enhanced sense of civic responsibility (Bringle and Hatcher 1996).

Service-Learning Project Implementation

At the beginning of each semester I introduce my ENV 110 students to two potential project paths. The first path is an online reflective lab journal that students complete after each unit (there are three units during the semester). The journals connect the field trip experiences with the content of the course (readings, class discussions, video reviews, and so on). The other path I offer to students is to become involved in a service-learning project by working with the UMF's Sustainable Campus Coalition (SCC). The SCC is an organization comprised of students, faculty, and staff that facilitates environmental activism and student civic engagement (SCC 2012). The organization was created as part of the university's commitment to building a sustainable future and is highly supported by the administration. The campus Green Vision Statement (see below) had already set us on a path toward educating the campus and community and modeling best practices in our own use of energy and resources.

UMF's Green Vision Statement

Sustainable practices will be an integral part of our campus management and operations, in building and renovating, in reducing pollution and waste, in using appropriate energy resources and materials, and in protecting and incorporating the native environment in our campus spaces. Indoor and outdoor environments should be healthful and aesthetically pleasing. Committing to environmental responsibility means that we will also set goals for ourselves and monitor our progress towards them.

The mission of the SCC has broadened considerably since it began in 2002 to include public education; collaborations with community organizations, municipalities, and schools; assessment and mitigation planning associated with greenhouse gas emissions; improvement of recycling on campus, development of a campus organic garden and orchard; encouragement of local food and institutional composting; and reduction of automobile idling. As one of the faculty coordinators for the SCC, I am in a position to provide my environmental science students with the list of projects that the SCC will be focusing on during each semester, and students are introduced by me to SCC members as they attend the beginning of the semester meeting. My role in this service-learning process is "to act as a facilitator for the project, set the learning objectives for how service learning fits into the curriculum, establish reflective opportunities for students, and assess the final products of the project" (McDonald and Dominguez 2008, p. 14). The student members of the SCC are paid by the SCC coordinators to initiate a variety of environmentally focused projects on the UMF campus. Their role is to be campus environmental project leaders. I pair one or two of my environmental science students with one or two SCC project leaders and this helps my students stay focused and motivated throughout the duration of their service project. My ENV students must attend at least six SCC meetings, where they choose a project that interests them and coordinate their efforts with SCC project leaders. Part one of the project involves students conducting background research and writing an essay discussing the importance of promoting environmental awareness on campus and how UMF compares with other Maine college campuses regarding environmental initiatives and progress. Part two of the project involves working on campus and documenting their efforts. I check in with each team every two weeks to make sure that my ENV students are participating and making progress. Part three occurs once the on-campus SCC project is completed. The students create a reflective video that chronicles their efforts and what they learned from the experience, and they present their video to my ENV 110 class and members of the SCC.

Both the unit lab journal reflections mentioned previously and the service-learning project are worth the same number of points. As I introduce these requirements, often someone asks, "Why would anyone choose the service-learning path instead of just doing the unit lab journal reflections—those seem easier?" I respond by saying that it all depends on how involved they want to be. As we explore numerous environmental issues and discuss the challenges we are facing as a species, individuals must decide whether or not they want to make a difference. The service-learning project provides that opportunity. I add that while there is nothing "wrong" in choosing the lab journal reflections, they need to realize that their engagement/involvement with the course material will be very different from someone who is actually doing what they would be writing about in a reflection. As with anything, choice is the key, and it depends on the desired level of involvement. "Service learning offers exciting opportunities for pedagogy that can enhance natural science education. However, historically service-learning has been underrepresented in natural sciences courses at institutions of higher education" (Curry, Heffner, and Warners 2002). I can understand why this might be the case. After attempting to incorporate service learning into this course over the last five years I have learned that it is critical in an introductory course with 40 students to give students a choice as to whether or not they want to participate in a service-learning project. During previous semesters, when service learning was required in my course, it took a tremendous amount of time and effort to ensure quality projects and consistent student participation throughout the semester. Not to mention, my antacid consumption during those semesters was beginning to impact the departmental budget. I had to find a better way! As I continued to modify these project offerings, I have become more efficient in outlining the specific criteria that need to be fulfilled, and as one of the coordinators to the SCC, I am now able to monitor student attendance and participation. During previous semesters the average number of students who chose a service-learning project was approximately 4–6 students (approximately 10–15%). The last three semesters, out of a class of 40 students, 25–30% chose to do a service-learning project. This entire process was daunting at first but as I narrowed the focus to involve only the UMF campus and as I facilitated student participation with the SCC, the projects have become much more involved and have had quite an impact on campus.

Evidence of Success

The projects that have been most successful thus far are those that have now been institutionalized—in essence, they are now a part of our campus culture and are completely supported by the university administration. Three projects will be discussed, each one of which involved students from my environmental science course and SCC project leaders.

Project #1: The Green's List Certifying Residence Halls

The idea to create the Green's List came about while developing UMF's Green Vision Statement. UMF has three U.S. Green Building Council (USGBC) Leadership in Energy and Environmental Design (LEED)-certified buildings and is committed to having all new construction be LEED-certified (USGBC 2012). However, like most universities, UMF has many old buildings and the idea at the foundation of this project was determining how to make these old buildings more sustainable. The thought of having our oldest residence halls becoming LEED-certified was the beginning of the Green's List development. This project has run every semester for the last two years. ENV and SCC students developed a green-certified checklist based on 10 categories with variable points per category. Students visit each room in the residence halls and survey residents regarding their energy usage, such as lighting, small appliances, computer use, and thermostat setting, as well as their recycling habits. They compile a score for each room and based on that score students rooms are certified as bronze, silver, gold, or best of all, green! Residents are then provided with energy-saving suggestions, and also local resources that they can access to become more sustainable. The success of this project was highlighted in the UMF alumni publication; below are brief excerpts from the students involved in the project as well as comments from the former UMF facilities management director regarding project impact.

- The goal of the project is to inform and encourage a sustainable lifestyle that works for the individual. It's the little things that add up to make a big impact on our energy savings here at UMF (Rachel Fritschy, SCC Project Leader).
- It's interesting to see how other people live, and sometimes it's reassuring to see that a lot of students on campus are saving energy because of the way their parents raised them. That's where the new generation is rising, and we're witnessing it right in front of our eyes. People are waiting for their [incandescent] light bulbs to burn out so they can put in the CFLs. It's good stuff! (Jourdan Merritt, SCC Project Leader).
- According to former facilities management director Robert Lampaa, "In fiscal year 2010, UMF reduced its energy consumption by 4% from the previous year, leading to savings of approximately $48,000. This is attributed to many factors, including upgrades to infrastructure and more energy-efficient improvements on campus, but I think that behavioral changes resulting from SCC efforts like the Green's List play a vital role. People are simply being more reasonable and responsible with energy use on campus" (Ottman, 2010).

Comments from ENV students demonstrate their excitement:

- The Green's list helped to create an awareness of sustainability. Other people my age, living in the dorms, may not be focused on the environment, recycling and living sustainably. The Green's list helped to promote conservation of energy and gave tips to my peers on how to live more sustainably. People my age should be concerned with these issues. One day, my generation will be the leaders of this country and it is important that we are all well educated on the damage that is being done to the environment.
- This service-learning project meant a lot to me and this community. I found that most students I surveyed answered the questions with great respect and were very curious as to how they could improve. If the students are better able to conserve energy, it could mean lower tuition, which would greatly help the community and students who struggle financially.

Project #2: UMF Trash Day

Each semester SCC and ENV students raise campus awareness regarding the amount of trash generated in the residence halls and how much of that trash could have been recycled. Students ask our facilities management workers to bring all the trash generated in the dorms within a 24-hour period and place all the bags on a large tarp in front of our library, which is the center of campus. Students put on rubber gloves and they go through each bag of trash, separating actual trash from recyclable materials. As university faculty, students, and staff walk past the library, groups of students explain what is happening and why it is so important to recycle on campus. Results from this past fall showed that roughly 22% of what students are throwing away in the residence halls could have been recycled. This was a notable decrease from past years, when the percentage of recyclable items thrown away was roughly 30–40%. There was a huge decrease in the amount of paper, cans, and bottles thrown away. UMF students seem to be more conscious of what they are tossing in the trash (SSC 2012). In addition to the ENV students participating in this event, several archaeology students also participated as part of a class project. The event was highly publicized around the state, as in this excerpt from the Lewiston Sun Journal:

> FARMINGTON—Archaeology students wearing safety gloves sorted through trash Wednesday to determine patterns of students living in dormitories at the University of Maine at Farmington. This was the annual UMF Trash Day. Trash collected over a 24-hour period from the residence halls was piled on the Mantor Green on South Street. The day is sponsored by the UMF Sustainable Campus Coalition, which was conducting its own project of trying to determine what percentage of the trash could have been recycled." (Perry 2011)

ENV students wrote the following regarding their participation during trash day:

- Attending the six meetings of the SCC, my partner and I were so surprised how committed they were in trying to get UMF students more involved in a variety of projects. It really motivated us to get more involved with these issues.

- My generation should be very concerned about this issue. We will be voting on where the next landfill will be in our state and the amount of trash that is going into our landfills is enormous and this just can't be good for our environment.

Project #3: Getting Local and Organic Foods on Campus

This project was first introduced by two of my ENV students during the Fall 2009 semester. They approached the UMF SCC with their idea. They wanted UMF's dining services to provide healthier local and organic foods. They started by interviewing the UMF Director of Food Services and his staff and worked around several challenges to provide a locally grown Thanksgiving feast for UMF students. The challenges they addressed included the inspection of local growers, availability and consistent supply, and cost. The students addressed those challenges by suggesting that our dining services staff start with one meal to determine whether or not it could be feasible for UMF to incorporate local and organic foods on a regular basis. The meal that everyone decided upon was the Thanksgiving feast. My ENV students researched and found local producers and coordinated food inspections with the dining services director. Once their work was completed, 75% of the meal came from regional and local growers, many from Farmington and surrounding towns. The Thanksgiving feast was a huge success and the students' dedicated efforts really started a trend on the UMF campus as our dining service continues to work toward providing more local and organic foods in the university cafeteria and snack bar. During this local Thanksgiving feast students interviewed numerous people to determine what they thought:

- "It helps the local economy, reinforces our connections with UMF and the local community and it's good" (UMF President Theodora Kalikow).
- "The food was terrific, I enjoyed the fact that it was from local Maine people, I like supporting Maine's economy" (UMF student).
- "Even though some of the food was more expensive, for example if UMF decided to get the turkey's that were provided for this meal from a commercialized farm the cost would have been $790.40, our cost for the local turkeys was $2,150. Yes, it is more expensive, but people will notice the tremendous difference in quality" (UMF Food Service Director).
- "It was made very clear, that students value local first and organic second. Our next focus will be on the snack bar. If students continue to show an interest we will do our best to meet that demand" (UMF Assistant Director).
- "It was great to have the opportunity to work on this type of project in our class. It was fun and we actually made a difference, so all of this work has really paid off" (ENV student).

The UMF dining service has changed considerably since this project was implemented. The following is directly from their website:

Sustainable Food: Local, Organic, Sustainably Certified, Natural, etc.

We understand the power and centrality of food in our daily lives and interactions, and recognize that our food choices have a significant impact on our health, culture, environment, and local and global economies. We are committed to fostering new connections from field to fork and changing the culture of food by nourishing our guests with menus that emphasize fresh whole foods that are raised, grown, harvested and produced locally and/or sustainably wherever possible, and prepared in ways that respect and maintain quality, freshness and pureness" (UMF Dining Services, 2011).

Next Steps

As with any new endeavor in teaching, at the start, the challenges appear to be more numerous than the benefits. However, as I became a better coordinator and facilitator of these projects, I realized that students really want to step up to those challenges and contribute to the campus community. By providing students with a choice of the type of project path they want to take (service learning vs. unit lab journal reflections), the students that choose the service-learning path do so because they really want to be involved in shaping how UMF achieves its sustainability goals. The students who choose to focus their efforts on the journals are those who feel they would be more successful making connections through this other avenue. As I continue to assess these projects and student learning gains, I would also like to devise a service-learning miniproject, where students participate in smaller-scale campus projects that can be completed within a couple of weeks rather than taking on a semesterlong project. The next steps in project and course development will be two forms of assessment. The first will be a survey of student behavioral changes regarding personal environmental practices comparing the level of change between the students who participate in a service-learning project versus the students who choose to complete the unit lab journal reflections. The second would be to gather data regarding student impacts on campus through the various service-learning initiatives implemented over time.

> Service learning is now a major national movement at every educational level, and is a particularly powerful force in undergraduate education. Connecting academic study with community service through structured reflection is widely recognized as contributing to learning that is deeper, longer-lasting, and more portable to new situations and circumstances. Campus Compact recently reported a three-fold increase in just four years in the number of full-time faculty teaching service-learning courses, from 14 per campus in 2000 to 40 per campus in 2004. (Ehrlich 2006)

This service-learning movement is not only supported by numerous national organizations such as Campus Compact and Science Education for New Civic Engagements and Responsibilities (SENCER), but many universities now have service-learning centers supporting dedicated groups of faculty. "College and universities are institutions that have major impacts on their communities through their use of resources, the physical area they cover, and through their modeling of environmental stewardship" (Curry, Heffner, and Warners 2002). By using environmental stewardship as a theme, perhaps a potential doorway for instructors of other natural sciences disciplines might be to create an environmentally themed unit related to that discipline. For example, one of my colleagues in chemistry who taught a unit on water quality had his students analyze the effluent from the local town sewage treatment plant. Students tested for the cleanliness of water by calculating how much dissolved oxygen exists in the water sample to begin with, then testing again after five to seven days. The greater the difference is, the higher the biological demand for oxygen and thus the more bacteria and other contaminants in the water. Students then shared those results with the treatment plant managers.

As I continue to explore various ways to implement service learning in my own course, it is very helpful to have these discussions with colleagues from a variety of science disciplines. This type of pedagogy will not necessarily work well in all science courses. It is up to the instructor to decide if the course goals match with this type of pedagogy. It is important to note that when trying this approach, or anything new, that the instructor should not "toss it" if it does not work well the first time. Based on my five years of experience, I can honestly say that the first three were very challenging. I believe I have finally developed a service-learning model that works well in fulfilling my course objectives and is also well received by the students who participate. However, there is still a lot of work to be done, especially regarding assessment and the effectiveness of this style of teaching.

Acknowledgments

I would like to thank Dr. Tom Lord for encouraging me to submit my work to this publication and also my heartfelt gratitude to an amazing mentor, colleague, and friend, Dr. Mary Schwanke, for helping me edit this chapter.

References

Bringle, R. G., and J. A. Hatcher. 1996. Implementing service learning in higher education. *Journal of Higher Education* 67 (2): 221–239.

Brown, F. 2011. Finding environmental education in the national science education standards. *Electronic Green Journal* 1 (14): 1–7.

Curry, J., G. Heffner, and D. Warners. 2002. Environmental service-learning: Social transformation through caring for a particular place. *Michigan Journal of Community Service Learning* 9 (1): 58–66.

Ehrlich, T. 2006. Service-learning in undergraduate education: Where is it going? *Carnegie Foundation for the Advancement of Teaching. www.carnegiefoundation.org/perspectives/service-learning-undergraduate-education-where-it-going*

Gutstein, J., M. Smith, and D. Manahan. 2006. A service-learning model for science education outreach. *Journal of College Science Teaching* 36 (1): 22–26.

Hungerford, H. R. 1996. The development of responsible environmental citizenship: A critical challenge. *Journal of Interpretation Research* 9 (1): 25–37.

Littledyke, M. 1997. Science education for environmental education: Primary teacher perspectives and practices. *British Educational Research Journal* 23 (5): 641–659.

McDonald, J., and L. Dominguez. 2008. Service learning: Taking action for the environment. *Michigan Science Teachers Association (MSTA) Journal* 53 (2): 11–17.

McKibben, B. 2007. *Deep economy: The wealth of communities and the durable future.* New York: Times Books.

National Research Council (NRC). 1996. *National science education standards.* Washington, DC: National Academies Press.

Orr, D. 1996. What is education for? Six myths about the foundations of modern education, and six new principles to replace them. *The Learning Revolution* (27): 52.

Ottman, K. 2011. Making a list, checking it twice. *Farmington First Alumni Magazine* 16–17.

Perry, D. M. *Lewiston Sun Journal*. 2011. Students examine dorm life through trash. October 13.

Shriberg, M. 2004. Sustainability in U.S. higher education: Organizational factors influencing campus environmental performance and leadership. Unpublished PhD diss., University of Michigan.

Sustainable Campus Coalition (SCC). 2012. Sustainable campus coalition. *http://sustainablecampus.umf.maine.edu*

UMF Dining Services. *www.campusdish.com/en-US/CSNE/UnivMaineFarmington*

U.S. Green Building Council (USGBC). *www.usgbc.org*

Assessing Student Attitudes and Behaviors in Interactive Video Conference Versus Face-to-Face Classes: An Instructor's Inquiry

Thayne L. Sweeten
Utah State University–Brigham City

Setting

As a land-grant university, Utah State University (USU) is charged with reaching out to educate students across the state of Utah. This large state spreads across a diverse geography from the densely populated valleys of the Rocky Mountains in the north to the beautiful deserts of the Great Basin in the west and the Colorado Plateau with five national parks in the south and east. The regional campus and distance education arm of the university comprises 27 distance campuses and delivery locations throughout the state. Nearly half of the 31,830 students enrolled in the university in 2011 received classes through the regional campus system, which has an outreach not only throughout the state, but throughout the country and world.

The first regional campus was established by the Utah legislature in 1967 and by 1968 professors would travel weekly by plane 300 miles roundtrip to teach courses at this site. In 1996 a statewide satellite system was installed, bringing students from multiple locations together. With the advent of the internet, USU students enrolled in fully online courses starting in 1997. Technology improvements brought about use of a two-way audio/two-way video system known as interactive video conferencing (IVC) in 2007. During the fall of 2011, 293 classes were delivered by interactive video conference (IVC). It was shortly after the installation of the IVC system, and following my sudden introduction to both IVC and university teaching, that I was hired as a new faculty member at USU's busiest regional campus and began using this technology to teach classes to students across the state.

My introduction to teaching was quite abrupt. Following completion of graduate school, I took a position as a postdoctoral research fellow at Utah State University attempting to understand biological causes of autism. Since I had been involved in research for some 10 years, I was relatively comfortable in this environment. Research design and analysis had been an ongoing part of this work, so when the call came from the university that they were in desperate need of an instructor for a course on research design and analysis, I offered to jump in. Indeed, it was a jump! The course had been running for a week when the previously scheduled instructor unexpectedly left. The students were PhD students in psychology and education, many of whom were accomplished professionals. As I entered the room to teach the class for the first time, there were no students present. I found myself in a "cockpit" with an array of cameras, controls and screens to enable the transmission of the class to various locations scattered throughout the state via IVC.

Focusing on National Science Education Standards

My new position provided a novel challenge that is faced by many instructors as we adjust to the exciting technological advances of the times. Distant education through means such as IVC is an emerging educational delivery system that many instructors have not had a history of experiencing as students. Most instructors today were students of traditional face-to-face classrooms and as such have a lifetime of experience to draw upon in the workings of a traditional classroom, but not for IVC settings. Given the relative novelty of this teaching approach, there is great need to understand how to best teach while using this technology.

This chapter describes my efforts to apply the National Science Education Standards (NSES) generally, but, in particular, Teaching Standard C inviting, "teachers of science (to) engage in ongoing assessments of their teaching... and (to) analyze assessment data to guide teaching" so as to better understand the IVC classroom.

As a participant at the National Academies Summer Institute on Undergraduate Education in Biology and through involvement at national teachers' conferences, I have been introduced to the concepts of inquiry and scientific teaching. The NSES describes inquiry as the students' "use [of] scientific reasoning and critical thinking to develop their understanding of science ... including asking questions, planning and conducting investigations, using appropriate tools and techniques to gather data, thinking critically and logically about relationships between evidence and explanations, constructing and analyzing alternative explanations and communicating scientific arguments." In a similar vein, scientific teaching involves bringing the rigor and spirit of science research to teaching, including the ongoing scientific assessments of an instructor's own methods.

Many of us who teach science have a background in scientific research and therefore inherently understand the power of inquiry in driving learning and the role of the scientific method in answering questions. The NSES state that "inquiry into authentic questions generated from student experience is the central strategy for teaching science." It could also be said that questions generated from a teacher's experience are a central force in shaping our teaching. As I began teaching IVC classes, many questions arose regarding the students' experiences in these classes. These questions provided opportunity to use inquiry and scientific teaching to provide the answers and understandings that guide teaching.

Features of the Instructional Setting

A bit of background is necessary before mentioning the question that guided my inquiry and search for answers. The courses that I teach include human anatomy, human physiology and elementary microbiology to a variety of majors many of whom are planning to apply to nursing programs. The setup of these courses is unique. A typical IVC class will have up to 50 students in a classroom at the origination site experiencing the class face-to-face while one or more distance sites will experience the course real-time over the IVC feed. A video camera and microphones capture the class at the origination site and deliver it to the distant site where the video feed is projected on a large screen in front of the receiving class. This screen will project an image of the instructor and using a split screen division can simultaneously project other teaching images such as PowerPoint slides, videos, or real-time images from a document camera, a modern digital version of an overhead projector. Audio is also delivered to a speaker system in the receiving room. As this is a two-way system, cameras and microphones also capture sight and sound from the receiving room and send it back to the origination room where the distant class can be viewed by the instructor on a monitor and comments and questions from the distant sites can be heard directly at the receiving sites. The distance sites receive the class with the same audio and visuals as the face-to-face class only lacking the physical presence of the instructor.

When delivering an evening course to multiple sites, both face-to-face and via IVC, there are a number of variables that need to be juggled by the instructor. One of those can be the different populations of students at the various sites. For instance, a course typically contains a face-to-face section with a number of mixed-major students recently out of high school combined with more mature students returning to school after time in the workforce. A distant site can include students from a state-run technical college studying to become a practical nurse while other distance sites can consist of students from a private college who are vying for a spot in that school's registered nursing program.

Implementation of the National Science Education Standards

After about a year of teaching such face-to-face/IVC courses, I started noticing that students in the face-to-face sections typically performed on average a grade better that the students at the distance sites. This observation sparked my scientific curiosity, and I began to ask questions and make inquiries as to why. I needed to consider a number of variables that could help explain the difference. It was apparent that the variable of instruction was held constant as the students were receiving the exact same course simultaneously at the different sites. Two other overreaching variables that could help explain the difference included differences in the student populations and differences in the mode of delivery at the various sites. While it could be hypothesized that both of these variables could be important factors, I decided to focus my assessment efforts on better understanding the role of the different modes of delivery. In particular, my inquiry focused on student attitudes and behavior toward the different deliveries. My rationale for focusing the assessment in this area was to better understand how the students felt and reacted to the two forms of delivery and therefore provide information on how to shape teaching to better meet their needs.

One thing that I have learned to appreciate as a laboratory scientist is the relatively tight control of variables in laboratory settings. In the sterile hoods of research labs, scientists control minute details, but this is not the case in the classroom "laboratory." Random sampling is often out of the question and the real world environment is difficult to control. However, as circumstances allowed, during one semester my teaching schedule afforded an excellent opportunity to gather some data in a relatively clean design.

This particular semester I was asked to teach a section of human anatomy to a group of students face-to-face on campus and simultaneously broadcast the class 100 miles away to another site. The class met twice a week. However, I was also asked to teach the cadaver lab for the distant site, so it was arranged that I would originate one class a week on campus and then travel to the distant site for the second class of the week to also teach the cadaver lab. Therefore, one day of the week I would originate from one site and broadcast to the other. For the second class this scenario would be reversed. This arrangement allowed for an excellent opportunity for students to experience an equal amount of face-to-face and IVC broadcast classes during the same semester from the same teacher.

To take advantage of this circumstance, a short survey was created to assess student behaviors and opinions toward the different modes of delivery. This survey was passed out to the students along with their final exam at the end of the semester (see Figure 15.1).

Figure 15.1. Opinion Survey Assessing Student Behavior and Attitudes Regarding Face-To-Face Versus Interactive Video Broadcast Delivery

Please **circle** the answer that best applies regarding the different types of lecture delivery, in person and broadcast via interactive video conference.

1. I am more likely to ask questions in class when the instructor is:
   a. Present in the classroom b. Broadcasting from a distance c. It makes no difference
2. I am more likely to answer questions asked by the instructor when the instructor is:
   a. Present in the classroom b. Broadcasting from a distance c. It makes no difference
3. I am more likely to attend class when the instructor is:
   a. Present in the classroom b. Broadcasting from a distance c. It makes no difference
4. I am more likely to be tardy or leave early when the instructor is:
   a. Present in the classroom b. Broadcasting from a distance c. It makes no difference
5. When the instructor is speaking, I am more likely to have personal conversations with fellow classmates when the instructor is:
   a. Present in the classroom b. Broadcasting from a distance c. It makes no difference
6. I am more likely to participate in small group discussions when the instructor is:
   a. Present in the classroom b. Broadcasting from a distance c. It makes no difference
7. Did technology issues deter you from understanding what was being presented in class?
   a. Never b. Rarely c. Moderately d. Frequently
8. I feel that I learn better when the instructor is:
   a. Present in the classroom b. Broadcasting from a distance c. It makes no difference

Please include any additional comments regarding live versus broadcast delivery below.

There were essentially the same number of students at each site, 35 at one site and 36 at the other. They all completed the questionnaire with all 71 students responding to each question. The following data describes the student responses to the survey questions.

The first two questions deal with the student's comfort level in interacting with the instructor through asking and answering questions.

Figure 15.2. Survey Question #1: Answering Questions

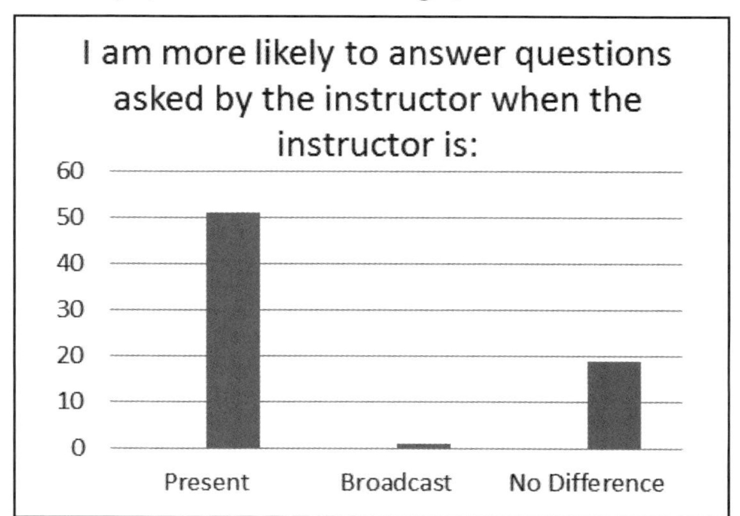

When the instructor was present, 56 of the 71 students (79%) were more likely to ask questions, whereas 15 (21%) said that it made no difference. No students indicated that they were more likely to ask questions when the course was broadcast.

Figure 15.3. Survey Question #2: Asking Questions

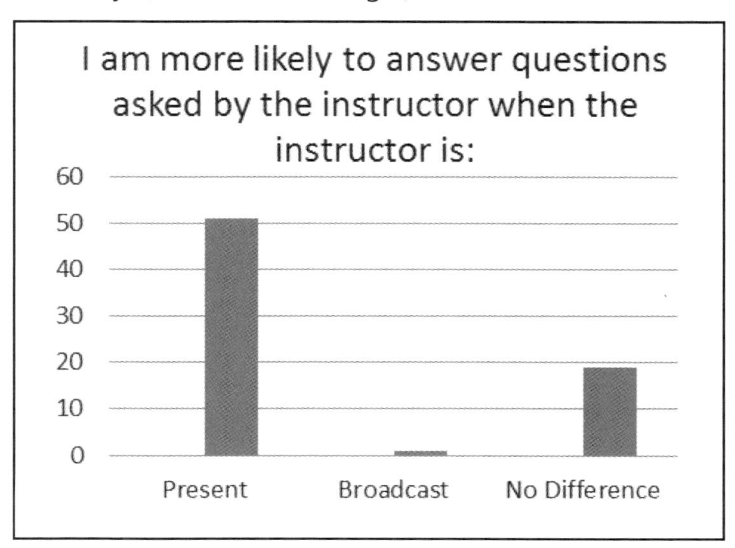

When the instructor was present, 51 students (72%) responded that they were more likely to answer questions, 19 students (27%) indicated that it made no difference, while 1 student (1%) preferred answering questions in a broadcast setting.

These two questions focus on an integral aspect of teaching: the ability to readily engage with students in asking and answering questions. In essence, the answers provided by the students to questions #1 and #2 indicate a reluctance to interact with the instructor when receiving a broadcast class compared to a face-to-face delivery. This finding leads to the follow-up questions as to why this is the case. Although there are no data in this regard, part of the reason may be due to some of the limitations of the technology. One of the imperfections of the two-way audio system is that if the microphones at a receive site are left on, there is feedback generated that is bothersome and over time causes the audio system to cut out. Therefore, the microphone at a receive site can only be left on for brief periods of time while students answer and ask questions. In order for the microphone to be turned on, students must either have access to an on/off switch (which is often not the case) or ask a classroom aide known as a facilitator to turn on the receive site microphone(s). This extra step in responding to questions or making comments can take time and is a technical deterrent to class participation.

Questions #3 (Fig. 15.4) and #4 (Fig. 15.5) address issues related to attendance.

Figure 15.4. Survey Question #3: Attendance

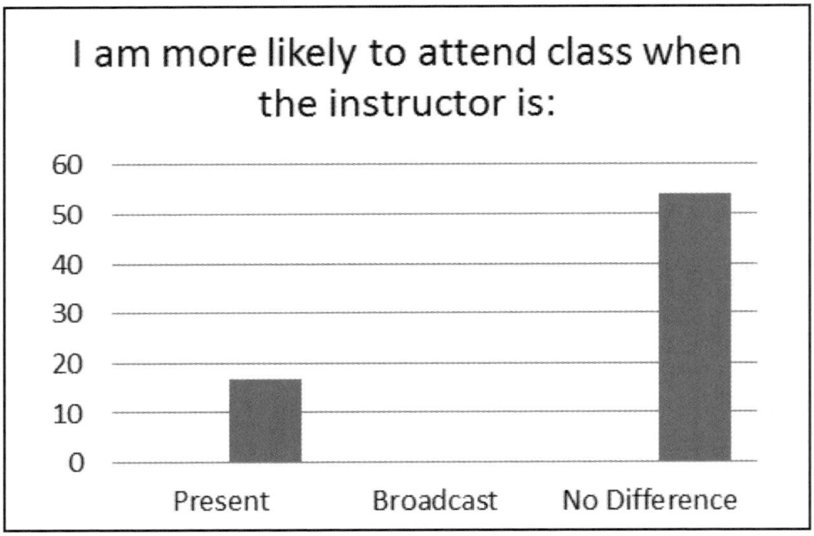

The majority of students (54 [76%]) responded that delivery made no difference in attendance, whereas 17 students (24%) indicated that they were more likely to attend when the instructor was present.

Figure 15.5. Survey Question #4: Other Attendance Issues

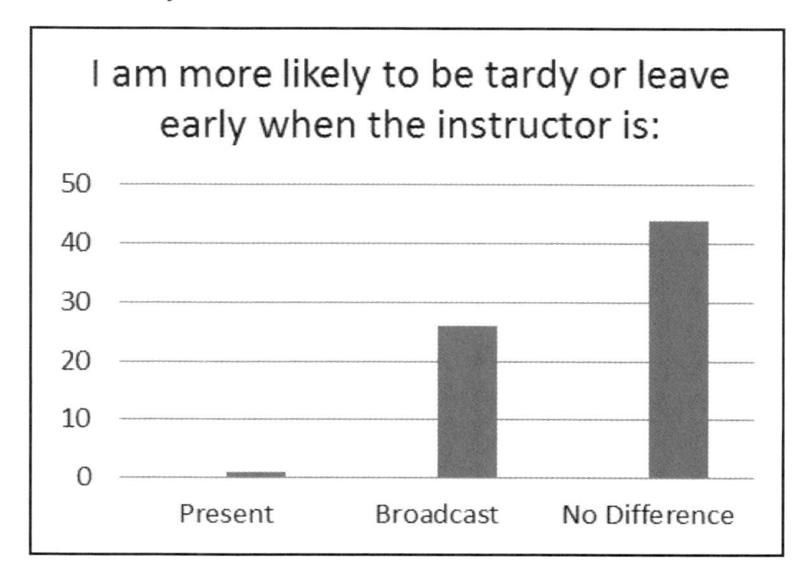

Of the 71 students, 26 (37%) indicated that they were more likely to be tardy or leave early in a broadcast class compared to 1 student (1%) in a face-to-face setting. For 44 students (62%) it made no difference.

The results of questions #3 and #4 reveal that students are less likely to attend a broadcast course and more likely to be tardy or leave early if they do. A factor that could partially explain this is the increased anonymity at a distant site. It is often difficult for instructors to clearly recognize the students at a distance site as their image may be small, unclear or perhaps outside of the view of the camera, thus making it easier to slip in or out unnoticed. Therefore, many of the social pressures that discourage such behavior in face-to-face classes may be reduced at a distance sites.

Question #5 (Fig. 15.6, p. 200) addresses disruptive behavior. While receiving a broadcast, 29 students (41%), compared to 0 face-to-face students, responded that they were more likely to engage in disruptive personal conversations. Forty-two students (59%) stated that delivery method made no difference in this regard.

These data validate a concern often expressed at broadcast sites that there is too much disruptive chatter occurring during class time. As the microphones are typically muted during class the students at the distance sites can talk to one another without the instructor hearing, and for many the temptation to talk out of turn can be great. An instructor can attempt to mitigate this by monitoring distance sites for visual cues of students conversing with one another, but this can be a challenging task to efficiently observe student chatter on a small monitor while teaching a face-to-face class and broadcasting to multiple sites at the same time.

One method of instruction involves splitting students up into small groups to try and solve problems or work together on various group tasks. Question # 6 (Fig. 15.7) addresses the amount of student engagement in these small groups.

Figure 15.6. Survey Question #5: Disruptive Personal Conversations

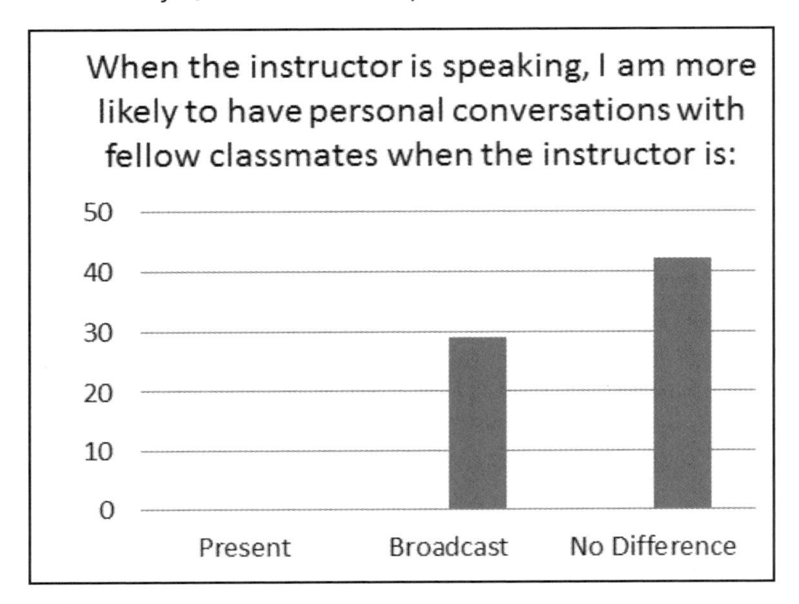

Figure 15.7. Survey Question #6: Small Group Participation

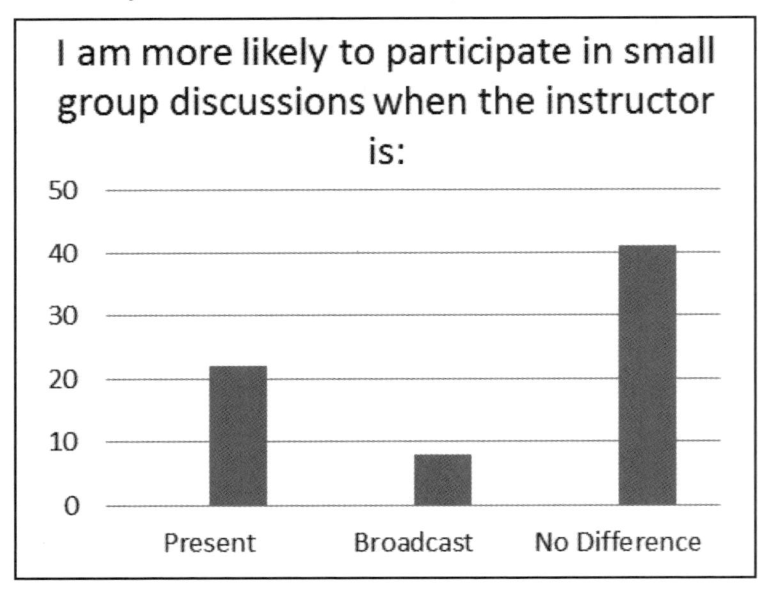

Responses indicated that 22 students (31%) were more likely to participate in small-group discussions when the instructor was present, while 8 (11%) stated they were more likely to participate when the course was broadcast. For 41 students (58%) delivery type made no difference.

Small-group work can be a powerful way for students to learn from one another while developing problem solving and expressive skills. It requires students to formulate thoughts and verbally use terms and concepts. In a face-to-face classroom an instructor can monitor the various small groups to provide guidance and insights as needed. This monitoring is not possible for students at a distant site; however, this teaching technique theoretically could be quite effective at distance sites, as students can learn from each other. Indeed, the above data does show that some students (11%) prefer these conversations at distant sites. Perhaps this is because they feel free to express ideas outside of the instructor's ear or perhaps the opportunity to engage other students in a face-to-face manner is a welcome change from watching the broadcast. On the other hand, 31% of the students were more likely to engage in these activities when the instructor was present. For these students the physical presence of the instructor provided more motivation to be engaged. The survey did not ask the students to explain the rational for their preferences. Determining such rationale could be an interesting subject for a future survey.

Question #7 (Fig. 15.8) asks about technological issues and whether they were a distraction from learning. As can be the case with complex systems, sometimes connections can be lost, sound can be unclear or buttons left unpressed. All of these events can interfere with effective communication and learning.

Figure 15.8. Survey Question #7: Technology Issues

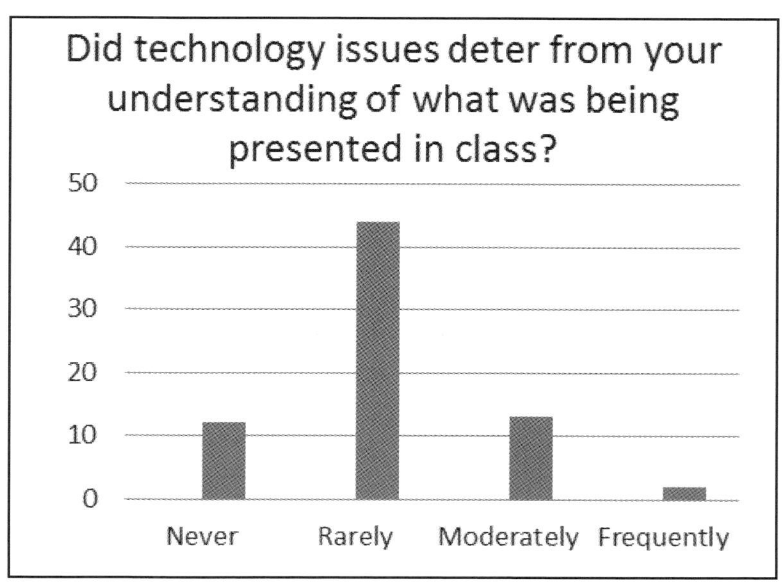

Of the 71 students, 12 (17%) claimed that technology issues never deterred them from understanding class materials while 44 students (62%) claimed rare interference from technology; 13 students (18%) were moderately and 2 (3%) were frequently deterred by technical issues.

The variability of the students' responses is interesting given that all the students were involved in the same course. For instance, 12 students never were deterred by technical issues while 2 were frequently deterred. This speaks to the idea that some students tolerated technical issues better than others. Overall, I was pleased with the finding that the majority of students were never or rarely distracted by the technology. In large part this is due to the quality of the technology and the technical support. For each broadcast course there is a facilitator in the origination room and each of the receive rooms. These facilitators work to mitigate any technical issues as they arise. Senior technical experts are also constantly available to address any concerns that cannot be corrected by the facilitators.

While it might be assumed that technical problems are primarily an issue only for the distance sites, this is not necessarily the case. If there is a technical issue at a receive site, the course is stopped for everyone until the problem is resolved. Also, microphones that are left on at distance sites can be disturbing to the face-to-face sections and some students just are not comfortable with "interruptions" from students in unseen places. Overall, I would agree that technical problems were rarely an issue. It has been my experience that as instructors, facilitators and others involved gain experience with the technology, the frequency of technical problems diminishes.

The last question, question #8 (Fig. 15.9), is a summary question that addresses the broad issue of overall learning. Do students feel like that there is a difference in how well they learn depending on the delivery type?

Figure 15.9. Survey Question #8: Overall Learning

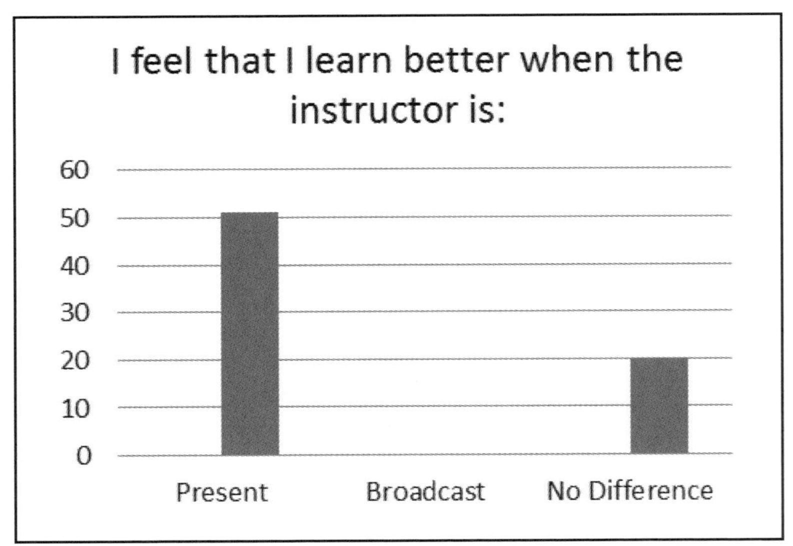

When the instructor was present, 51 students (72%) felt that they learned better, while 20 students (28%) felt that delivery type made no difference. No student chose broadcast as their preferred learning medium.

While it is notable that no student felt that they learned better in a broadcast section over face-to-face delivery, it must be kept in mind that these questions elicited subjective responses and do not empirically measure student behaviors at the various sites, neither do they measure student learning. Nevertheless, the responses to all of the questions are consistent with my anecdotal observations over the years and provide an informative assessment of student behavior and opinion.

The final item on the survey allowed the students to make any comments regarding their experience with live versus broadcast courses. As questions #1 through #8 were structured questions with limited choices for responses, this final open-ended assessment provided an opportunity to gain further insights. Some of the student comments regarding live versus broadcast courses were as follows:

- "I feel more comfortable asking questions or making comments when an instructor is present in the class, but I feel like I still learn just as well when he is broadcasting. "
- "It is easier to listen and understand when you are present in the classroom."
- "You bring such energy to class; it was always nice to participate in person... There were a few times that I thought of a question after class, that I would have asked if you were here, but forgot by the next lecture."
- "When the teacher is present, it sure is [easier] to interact with him, but as far as learning the class material, live and broadcast is the same."

The results of this survey indicate that when students are in a face-to-face class they are more likely to behave in a manner typically considered more conducive to learning. In turn, a majority of students reported that they felt like they learned better in a face-to-face class. These results must be interpreted in context. First off, the study was designed to better understand the attitude and behaviors of students at distance site so that the instructor could improve the quality of instruction provided to the distance sites. The insights gained from this study will guide improvements in teaching techniques to IVC sites and hopefully improve student experience in future semesters. Second, IVC delivery is a tool designed to most realistically reproduce a classroom experience for students separated by distance from sources of academic instruction. As such, IVC is a valuable tool for students who otherwise would have difficulty accessing educational opportunities.

The current data is informative for improving instruction. It is apparent that the technology does present a barrier to student participation in the class. With this in mind, I have made extra efforts when teaching duel delivery classes to encourage involvement from distance sites. As the face-to-face students are more likely to respond to questions during class, I have made it a point to frequently address questions directly to students at the distance sites. On certain days during the semester, I will declare that it is a certain distance sites "day," giving them the first opportunity to respond to all questions asked that day. Improvement in distance site room setup and ease of microphone access can also facilitate more student involvement. I have found it helpful to invite faculty colleagues to observe a class at a distant site to give feedback directly relating to how instruction translates to the site, as well as provide feedback on the effectiveness of receive room equipment and room configuration. On-site facilitators are also good resources for providing this kind of feedback and can also be tasked with helping control distractive chatter among students.

Various tools could be used to increase attendance of students at distant sites. One I have used is daily quizzes, typically administered at the beginning of class to encourage prompt arrival. On some days I have administered them toward the end of class. These quizzes also serve to prepare students for exams and encourage them to keep current with regular study of the material to avoid reliance on last-minute cramming.

To facilitate distant site participation in small-group discussions, I openly encourage those at distant sites to participate in these small groups and frequently make a point to ask students from distance sites to be the first to share with the class what they have discussed. The preference to the distance sites does not take away from the face-to-face class as during the discussion time I had already spent time monitoring their groups and interacting with them.

Live polling questions is another method that can be used to keep distant students engaged. The technology allows questions to be posted in a PowerPoint slide and then students can respond from face-to-face or distance sites either using cell phones or "clickers." The responses are displayed to the class in real time. This technology is effective for all students regardless of location. I>clicker and Poll Everywhere are two popular platforms that I have used.

The purpose of the survey was to better understand how students responded to IVC classes so that improvements could be made in teaching using this technology. To this end the survey did provide a greater understanding of the students' behaviors and opinions of IVC classes and it did provide input for informed adjustments to teaching style. However, this survey is just an initial step in the continual process of assessing and improving teaching.

Next Steps

What could be possible next steps in the assessment of delivery methods? While the survey was successful in determining behaviors and opinions, it also raises further questions, questions of why. Why were students more likely to participate in small-group discussions when the instructor was present? Why did students feel like they learned better when the instructor was present? It might be assumed that the physical presence of the instructor allowed more ease in interactions or added a "social" pressure to behavior in "appropriate" ways. Perhaps nonverbal communications that contribute to communication and thereby learning are difficult to transmit via IVC; however, the answers to these questions are not provided by the data and further assessments that specifically address these why questions could provide more direction on improving instruction in these dual-delivery classes. As well, studies that measure learning gains for IVC students who are exposed to different teaching techniques would also be informative in determining what styles of teaching are most effective for this medium. Finally, comparative studies that measure actual learning gains of students in face-to-face sections compared to IVC sites could be informative.

Other Investigations

As revealed in the literature, others have sought to assess and address the challenges of teaching via IVC. Dogget (2008) conducted a student survey on a class with both a face-to-face and video conferencing section. Due to space and budgetary constraints both sections were located on the same campus. Of the students in the course, 80% agreed that they would have been more comfortable in a normal class setting and 57% agreed that the videoconferencing was a barrier to their interaction with the instructor. Some of this dissatisfaction could be explained by the fact that students did not expect a video conference course when they registered and they were not separated from the instructor by a large distance as is the case with most broadcast courses. Dogget concluded that "given a choice, students prefer face-to-face interactions with an instructor," but "for off-campus learning, this technology has good potential" (p. 42). Knipe and Lee (2002) compared self-report research diaries from 29 students at a face-to-face site compared to 17 students receiving the same course via IVC. The students were Master's degree students taking a course on computer-based learning. Analysis of the diaries revealed that the IVC students "did not experience the same quality of teaching and learning as face-to-face students" (p. 210). Another investigation compared IVC versus face-to-face instruction for delivering lectures to large numbers of students on two different campuses. The class was taught face-to-face and simultaneously broadcast to a distance site. Each week the instructor would rotate the origination site so that students experienced both face-to-face and IVC delivery. Feedback systematically gathered from the students revealed that nearly half felt that receiving a broadcast was a disadvantage compared to face-to-face instruction (Freeman 1998). Similar sentiments were expressed in a survey of students who received instruction primarily via IVC but occasionally had the instructor originate from their site. Students indicated that having the instructor present at their own site was the optimal arrangement (Gillies 2008). However, not all studies report student dissatisfaction with IVC. Zerr and Pulcher (2008) found that participants in a one-day senior nurse leadership assessment day, designed for outside assessors to evaluate senior nursing students, were satisfied with the use of interactive videoconferencing. Some advantages of IVC assessment were reduced need for travel of the assessors and a reduction in anxiety for students being assessed.

IVC as a mode of course delivery has its advantages and disadvantages. The pros and cons have been well laid out in the literature (Townes-Young and Ewing 2005; Omatseye 1999). IVC's primary advantages can be summarized as being a relatively cost-effective way of bringing instruction to students at remote places in the most interactive way possible. Challenges include difficulty in "connecting" with students and thus maintaining their focus and interest.

Overall, the effectiveness of IVC instruction can be dependent upon the context in which it is used and the teaching methods applied. For instance, sustaining the interest of remote learners can be particularly challenging for exclusive teacher-centered teaching methodologies relying on long lecture sessions (Martin 2005). However, video conferencing is recommended by some as a good medium for classes that frequently use group work (MacIntosh 2001). Other researchers agree that interactivity, brisk pace, and a variety of delivery strategies are most effective for this media, with short 10–15 minute lecture sections interspersed with discussion, media presentations, question and answer, activities, and dialogic exercises (Martin 2005; Omatseye 1999).

References

Dogget, A. M. 2008. The videoconferencing classroom: What do students think? *Journal of Industrial Teacher Education* 44 (4): 29–44.

Freeman, M. 1998. Videoconferencing: A solution to the multi-campus large classes problem? *British Journal of Educational Technology* 29 (3): 197–210.

Gillies, D. 2008. Student perspectives on videoconferencing in teacher education at a distance. *Distance Education* 29 (1): 107–118.

Knipe, D., and M. Lee. 2002. The quality of teaching and learning via videoconferencing. *British Journal of Educational Technology* 33 (3): 301–311.

MacIntosh, J. 2001. Learner concerns and teaching strategies for video-conferencing. *The Journal of Continuing Education in Nursing* 32 (6): 260–265.

Martin, M. 2005. Seeing is believing: The role of videoconferencing in distance learning. *British Journal of Educational Technology* 36 (3): 397–405.

Omatseye, J. N. 1999. Teaching through teleconferencing: Some curriculum challenges. *College Student Journal* 33 (3): 346–353.

Townes-Young, K. L., and V. R. Ewing. 2005. Creating a global classroom. *T.H.E. Journal* 33 (4): 43–45.

Zerr, D. M., and K. L. Pulcher. 2008. Using interactive video technology in nursing education: A pilot study. *Journal of Nursing Education* 47 (2): 87–91.

Graduate Distance Education Programs in Forensic Sciences, Pharmaceutical Chemistry, and Clinical Toxicology for Working Professionals: An Evolving Concept With Practical Applications

Oliver Grundmann
University of Florida, Gainesville

David G. Lebow
HyLighter LLC

Setting

This chapter summarizes the graduate distance education programs in forensic sciences, pharmaceutical chemistry, and clinical toxicology at the University of Florida with a focus on the purpose and initial development of the programs, the growth of the programs over the past decade, and the integration of a new technology—HyLighter—to bridge the knowledge transfer gap that many science-based programs may encounter.

The past decade has seen significant changes in the way technology can be integrated and applied to science education on all levels. With a growing integration of technology into everyday tasks, students become more advanced in applying technology in the classroom. In an education setting, technology is already being used in the form of multimedia presentations that use readily available resources in both basic and advanced courses throughout all educational levels.

An important area of higher education is to provide support to working professionals who often have to maintain a balance between their work and private circumstances. This position presents specific challenges since those who have already obtained an undergraduate degree and entered the workforce often are reluctant or cannot afford to return to a university campus to continue their graduate education without significant consequences regarding their employment, financial, and private situation.

While distance education has emerged as an equally effective and efficient program of study in higher education, it has remained a major focus in the social sciences while the natural sciences—with the exception of a few institutions and organizations—have been reluctant to embrace the concept of distance education as a feasible method of education and knowledge transfer. This chapter examines the application of emerging technologies for online teaching and learning that deeply engages students with each other and the content to be learned toward improvements in learning outcomes.

Program Overview

The intention of all educational efforts is to provide students and learners with knowledge that can be applied (i.e., transferred) to their respective professional careers as well as providing a solid foundation to all graduates (from high school through to graduate school) to be productive and self-sufficient members of society.

The goals of the National Science Education Standards (NSES) are intended to "define a scientifically literate society" (NRC 1996). The major pathways to achieving these goals are to provide students with the knowledge and understanding of natural processes that can help them make educated personal decisions using scientific processes and principles. Furthermore, students should be able to engage in public discourse and apply their knowledge as productive members of society in their respective careers. Although these standards were developed for the K–12 setting, they continue to apply to higher education where more targeted and specialized knowledge that enables the learner to advance in their professional career is necessary.

Distance education has increased significantly in the past years, with 32% of all 2- and 4-year institutions of higher education reporting to offer degree or certificate programs entirely online (Parsad and Lewsi 2008). The report by the U.S. Department of Education also highlighted the use of asynchronous internet-based technologies as the primary delivery method for content and communication in online courses. The primary reason for offering online courses is to meet the increasing demand of students for a flexible schedule. In addition, learning approaches that are more student-centered and provide opportunities for independent study rather than lectures are important advantages that are being fostered in a distance education setting (see Figure 16.1). With the recent emergence of social networking, communication and interaction among students and with faculty has significantly shifted from the isolated classroom into a more diverse and integrative virtual realm. While distance education is often more writing intense, it also provides the opportunity for students to improve and receive immediate and individualized feedback from instructors on their performance that is distinct from a large traditional classroom lecture setting. Distance education is now able to provide students and working professionals alike with the opportunity to access information and receive credit. At the same time, students are able to advance their education throughout life with an increased level of flexibility that cannot be accommodated by the traditional classroom setting.

Figure 16.1. The Connections in the Online Learning Environment

The main purpose of the graduate distance education programs in forensic sciences at the University of Florida is to help learners who already have acquired a broad and fundamental understanding of natural processes through their prior undergraduate education and work experience to advance further. Graduate classes are distinctly different from K–16 coursework in that learners already have reached a level of professionalism and understanding that enables them to enter the workforce and be productive. However, as the complexity of the workforce becomes more specialized, additional and more targeted knowledge is often required. Learners in a graduate education setting are often self-motivated (Roper 2007) to (a) acquire new knowledge that directly translates into their current work environment, (b) refocus their careers on a different area, or (c) acquire supplemental information and skills that advance their performance as it applies to their profession. The primary focus of the distance education programs is therefore targeting working professionals to enable them to advance their careers and knowledge in a specific area applicable to their current or future career goals.

The programs are administered through the college of pharmacy at the University of Florida and include several full-time academic faculty with a wide range of expertise in the areas being taught. Additional adjunct faculty members supplement the full-time faculty to deal with specific courses related to their areas of expertise. Faculty is often supported by teaching assistants who add to the knowledge base in the topic area to assist the instructor and provide students with an up-to-date skill set to enhance their knowledge on targeted subject matter. Most teaching assistants have taken courses with the program themselves and work within the same area that they teach. Of note, faculty members of the distance education programs are located around the world—across the United States, Australia, South America, and Europe—providing a global perspective and views on topics that may differ between countries (e.g., legal setting in forensic sciences, best practice guidelines for treatment of a poisoned patient, and firsthand insights of developments in the pharmaceutical industry or variations in regulations).

Major Features of the Instructional Program

History and Development of the UF Distance Program in Forensic Sciences

The basic rationale that triggered the development of the distance education courses in forensic sciences was to provide working professionals with an opportunity to advance their careers and education while maintaining their current work-life balance. Educators have developed several approaches over the past decades to develop content for a distance education setting (including correspondence studies, video lectures, DVDs and CDs, and web-based delivery), which requires a balance between achieving the best learning outcomes for students while also allowing flexibility for faculty when content has to be updated. A suitable model for working professionals should allow for a high degree of independent study and deliver the content in written format as separate modules that are released over the course of a semester to students. Based on former experiences, providing online courses to international and national students in written format has been determined to best fit the working professional, who is the main target of the educational initiative. Delivery via videotaped lectures remains a challenge because of technological issues, especially in countries that are not as technologically advanced. Before establishing a full graduate program, the most feasible option was to implement single courses and evaluate student outcomes.

The first courses were offered in the fall of 2000 and evolved rapidly into the first graduate certificate and Masters program in Forensic Toxicology through the College of Veterinary Pharmacy under the guidance of Dr. Ian Tebbett, now Director of the Forensic Sciences distance education programs at the University of Florida. Further addition of courses led to the establishment of several new concentrations in forensic drug chemistry, forensic DNA and serology, and finally an MS in forensic sciences through the college of pharmacy. The general certificate and MS in forensic sciences allows students to choose 14 credit hours from among over 25 elective courses in addition to the required 18 credit hours of coursework. There are other programs at the University of Florida that also provide online access to their courses, such as the college of engineering, business administration, and education (for complete list please go to *www.distance.ufl.edu*). Each program is administered separately and thereby differs in delivery of content and organizational structure.

In recent years, additional programs in pharmaceutical chemistry and clinical toxicology have been established. Currently, a total of six different MS concentrations, two professional science masters (PSM) in forensic sciences and pharmaceutical chemistry, and eight graduate certificates are available to working professionals from around the world. While most of these courses are taught by faculty at the University of Florida, several courses in forensic medicine and environmental forensics are offered through partner faculties at the University of Canberra in Australia and the University of Edinburgh in Scotland. This provides a unique perspective in several regards: (1) experts in their area can contribute to an international program without having to relocate or students having to plan for travel and accommodations abroad, (2) a diverse faculty across the globe can provide various perspectives on current developments at a national and international level (e.g., legal proceedings that relate to forensic sciences or regulatory agency perspectives in pharmaceutical chemistry), and (3) the collaborative nature of the faculty and, therefore, between the programs increases visibility of the programs around the world and beyond the national level alone.

This collaborative approach has enabled the distance programs to expand and provide working professionals with flexibility and opportunity to continue their education without suspending their current work and life balance. For faculty involved with the programs, it has provided new opportunities to integrate a variety of tools to engage students and work on a flexible basis with students from across the world. Since most students and other faculty are spread across the globe and different time zones, faculty have to provide flexible learning tools and a study schedule that is able to accommodate working professionals.

The major platform for defining content is the open-source collaboration and learning environment (CLE) Sakai. Sakai was initially developed as a collaborative tool in 2004 by the University of Michigan, Stanford, Massachusetts Institute of Technology (MIT), and Indiana University. Since its inception with the intention to integrate and synchronize various instructional tools into an open source platform, the Sakai Foundation now has over 100 members from academia, industry, and contributing individuals that ensure a constant development and optimization of the platform. The major features of the Sakai platform are an individual website with access rights to respective course websites; discussion forums for exchange of ideas and posting of important information; an assessment tool for online timed quizzes and tests; an assignment submission tool for essay assignment submission and commenting by the instructor to provide individualized feedbacks to students; and a grade book for posting individual grades that also allows direct recording of points gained from online assessments, assignments, and graded discussion boards. Sakai was adopted in 2010 by the University of Florida, since it is an open-source platform that provides for a higher degree of flexibility in choosing which components are required and allows for significantly increased control over finances compared to commercial products such as Blackboard or WebCT. However, certain components that are not yet accessible via open source remain with commercial licenses.

In addition to Sakai for regular and frequent communications and submission of assessments, several courses use additional tools to increase communication and collaboration skills among students. Blackboard Collaborate (formerly called Elluminate) is a platform that allows for synchronous chat using voice-over-the-internet protocol (VoIP) combined with the use of an interactive virtual whiteboard that permits instructors and students to upload and share PowerPoint presentations and virtually interact with each other. Several courses within the pharmaceutical chemistry program have frequently used chat sessions. Synchronous chat sessions have been well received by both students and instructors and increase the understanding of the content as well as providing a feedback mechanism to the instructor about potential knowledge gaps that need further clarification. Recording the chat sessions allows students who were not able to attend to access the information and communicate with the instructor if questions remain. One drawback of the synchronous interaction is the time zone difference between students attending from across the world. This limits the collaborative aspect of this tool.

Two other tools have been implemented and tested over the past semesters starting in summer 2010. The first is a social annotation platform called HyLighter that supports online discourse and deliberation tied to highlighted sections of documents.[1] HyLighter allows students and instructors to highlight arbitrary (i.e., self-selected) parts or fragments of a multipage, multimedia (e.g., text, graphics, figures, tables) collection of documents and add comments to the margins of the pages. For example, an instructor asks questions related to highlighted fragments to assess knowledge and understanding of individual students. Alternatively, members of a class add comments to a document, evaluate their observations and responses to the material or problems posed, discuss potential pitfalls, and come to conclusions. The instructor can provide guidance at any point by replying to the comments of students. In sum, HyLighter combines highlighting and discussion threads to provide a versatile application for higher education.

The second tool that has been introduced recently for pilot-testing in some forensic science courses is CrimeSeen 360.[2] Crime scene evaluation is a central part of the forensic investigation and is a challenging task to communicate to students in a distance education setting. Students can use a web application or Apple's iPod, iPad, or iPhone to access complete photographs and walk-throughs of crime scenes. The photographs are taken with a fish-eye lens and allow a 360 degree round view of a room if centered correctly. Furthermore, the student can zoom in to evaluate and snap pictures of specific evidence to be collected. The snapshots can then be sent to the instructor with comments (see Figure 16.2). A similar application built by the same developer (TourWrist) is also being used by law enforcement officials for evidence location as well as for documentation purposes. This may provide a useful tool for crime scene documentation in general and, especially, for visualization of a crime scene in front of the jury and the public during trial proceedings.

1. HyLighter is a product of HyLighter LLC developed in conjunction with the University of Florida. More about the platform can be found at *www.hylighter.com*

2. CrimeSeen 360 is a product that was developed by TourWrist in conjunction with the University of Florida. More about the program can be found at *www.tourwrist.com*

Figure 16.2. Screen Shot of CrimeSeen 360 With Comment Box for Submission to Instructor

The "Working" HyLighter

Even as faculty and students increasingly adopt social tools to enhance collaboration and improve learning outcomes, a critical gap remains in information technology (IT) to support document-centered deliberation and sense-making practices. Although many applications available today support document-centered collaboration, two major deficiencies exist:

- How to collaborate on documents (e.g., Word, PDF, PowerPoint, Excel, HTML, and image files) that require many people (e.g., 10–100 or more students in a distance learning class) to share their thinking in a deliberative, controlled, organized, efficient, and auditable manner.
- How to properly source, disseminate, and make sense of many documents related to a topic of interest and filled with unstructured data (e.g., free-form text and graphics) in a coordinated effort.

One of the most widely used strategies for gaining understanding from text is to annotate or mark arbitrary or self-selected fragments with a highlighter pen and add relevant comments to the margins of the page. A new software product, HyLighter, supports a collaborative version of this activity that enables online classes of several to 100 or more to share their annotations and engage in discussions and deliberation. Instructor and students highlight fragments, add relevant comments in the margins, and create links between related fragments within and between documents as well as to other sources located on the internet.

Comments that are associated with each hotspot have provisions for threaded conversations and linking to related fragments in other documents (e.g., a user links a sentence that makes a claim in one document to a sentence that provides evidence for the claim in another). The approach enables large online groups to engage in discussions tied to specific sections of a document in an organized and efficient manner and make sense of information spread across many documents and various file types.

Figure 16.3 shows a document displayed in a browser with the HyLighter menu bar. Figure 16.4 illustrates how users link fragments between documents and different file types.

Figure 16.3. The HyLighter Screen

The browser displays the HyLighter tool bar below the browser tabs. The screen is split into two panels. The right panel holds the source with various color-coded sections. The left panel shows comments and related information submitted by contributors.

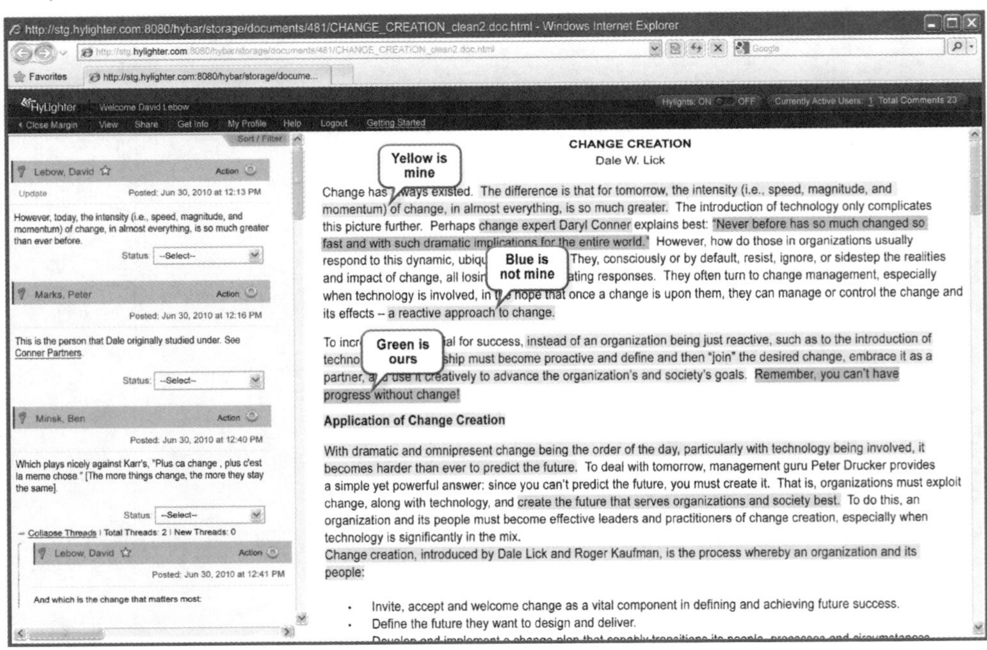

HyLighter provides a mechanism that enables users to (1) link related fragments between documents and comment on the relationship between them and (2) send a link to a specific fragment through e-mail and add a comment to the margin of a document by simply replying to the e-mail. These functions are accomplished by assigning a unique URL to each user-generated comment and related fragment, referred to as a comment-level URL (CLU).

Figure 16.4. Link-Related Fragments Between Documents

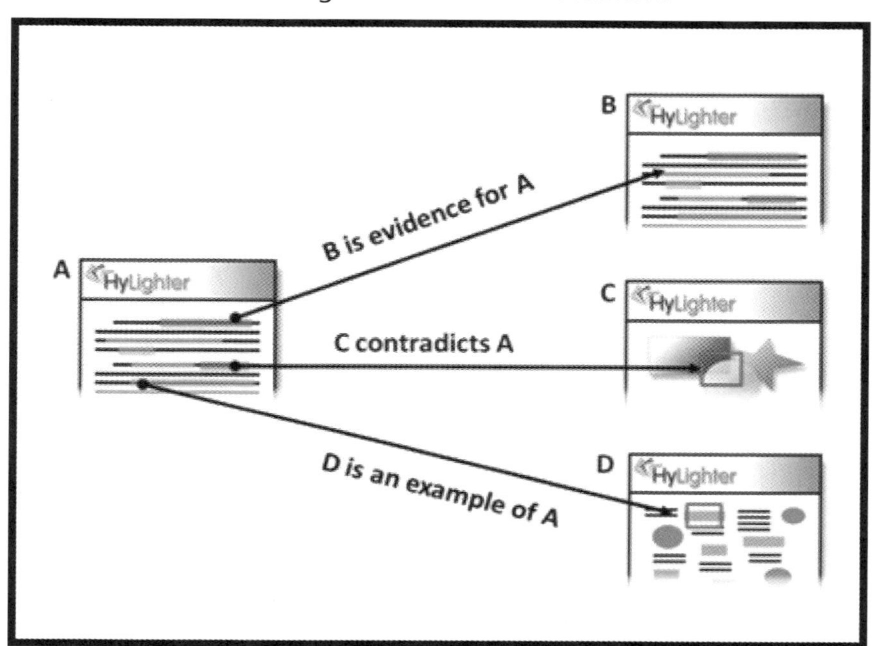

The features of a new technology are not what make the difference in desired learning outcomes but, rather, how the technology is used to support instruction and learning. Researchers in the learning sciences have learned a great deal about the development of expertise and expert performance, characteristics of expert learners, the qualities of effective teachers, ways to facilitate transfer of learning, basic principles for the design of learning environments, and application of information technology for instructional purposes.

The design of HyLighter embodies certain conjectures about learning and social context drawn from the existing research and theory base of the learning sciences. For example, this includes social constructivism (Lave and Wenger 1991), the theory of expert performance (Ericsson, Krampe, and Tesch-Römer 1993) and knowledge-building communities (Scardamalia et al. 1989).

From the broad perspective of the learning sciences, HyLighter enables three key functions:

- Makes the thinking of teachers and students that are ordinarily hidden become transparent and easily accessible for sharing with others, self-reflection, and feedback.
- Allows users to continuously compare their developing understanding to others, assess performance, and monitor progress.

- Supports sense-making efforts to organize, integrate, and synthesize information from multiple sources and perspectives.

Application of HyLighter in Graduate Science Courses

HyLighter was used to evaluate student knowledge after completion of the learning material in the graduate distance education course PHA5433 Fundamentals of Medicinal Chemistry. The HyLighter assignment has been used in this class since the summer of 2010 for five consecutive terms (summer and fall 2010, spring, summer, and fall 2011). During this time, over 100 students have taken the course and completed the assignment. The HyLighter assignment provided students the opportunity to transfer and apply the theoretical content learned during the course into a practical example. The course introduces basic concepts used in drug development such as physicochemical properties of drugs, degree of ionization of functional groups, solubility, principles of pharmacokinetics and pharmacodynamics, and metabolism. The asynchronous approach to this exercise allows students to first provide their own solution to a given problem and then discuss their answers later on among each other. Similar approaches have been evaluated for online medicinal chemistry courses offered to pharmacy students (Alsharif and Galt 2008).

The HyLighter assignment was provided to students as the last mandatory assignment after all other individual assignments and quizzes had been completed. The instructional material including graphs, text, and chemical structures was prepared as PowerPoint slides that were directly uploaded to the HyLigher server. As mentioned, initially students could see the questions and only their own individual answers until a certain deadline. After the deadline, they could see all comments, which allowed for a collaborative approach in discussing and agreeing on the best answers as a group.

Students were given access to the assignment, and the instructor and other students posted questions related to structures and concepts that students had to comment on. The instructor provided initial instructions on how to use HyLighter and its tools. Students could then highlight text and specific areas on the PowerPoint slides or post comments to parts highlighted by the instructor and other students. The interactive nature of HyLighter was apparent once students were able to read and discuss each other's comments. This allowed for both constructive discussions of concepts that were part of former lessons as well as acknowledgment if a student had made a miscue. The instructor also encouraged students to provide reasons for their initial answer choices and how their understanding changed after the instructor made all comments accessible (by changing the permission settings in HyLighter). Furthermore, students were provided with feedback by the instructor and were able to discuss the material during synchronous chat sessions after completion of the HyLighter assignment. Figure 16.5 shows a slide from the fundamentals of medicinal chemistry activity.

Figure 16.5. A PowerPoint Slide in HyLighter

The instructor has added two questions linked to different sections of the graphic. Students independently respond to the questions and are able to see the responses of their peers and the instructor when the instructor changes HyLighter permissions.

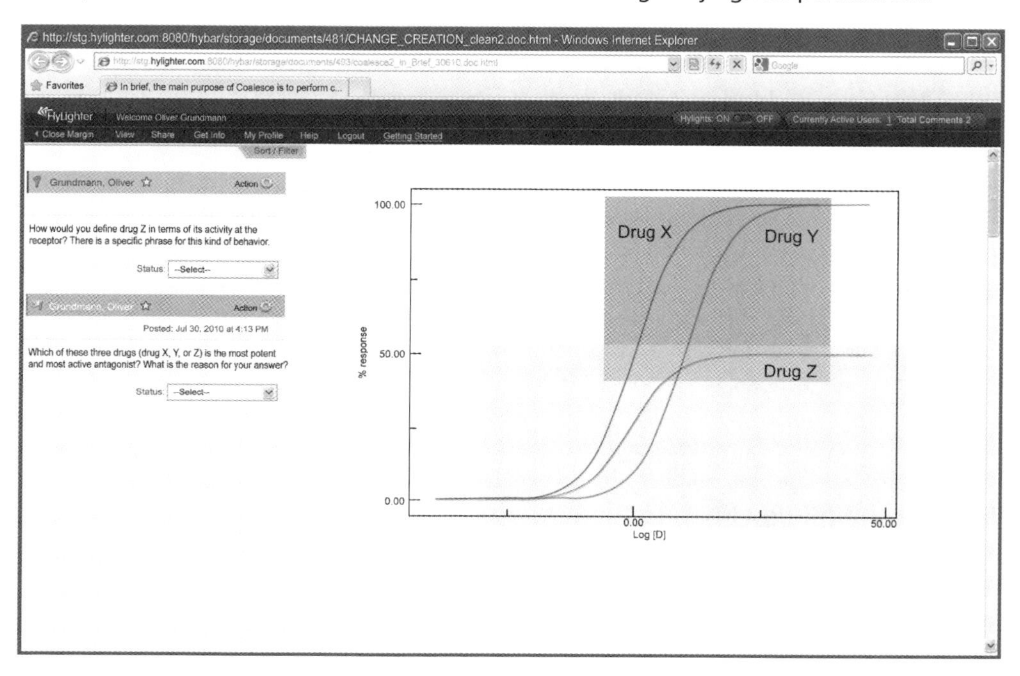

During chat sessions and the course survey, students provided positive feedback about the HyLighter assignment including the following comments:

- "I thought the Hylighter assignment was very useful, but not introductory."
- "I like the Hylighter assignment. Felt more like a discussion led module rather than just book reading. Liked reading everyone else's comments."
- "Interactivity is a plus plus."
- "I actually enjoyed the Hylighter assignment better than the others."
- "And you get to see when you don't know what you're talking about."
- "Yes, this assignment was really for me and made me search and go deep into the matter."

Students were able to apply acquired knowledge and experience and see the relevance of their new knowledge applied to the drug development process. The HyLighter assignment also helped in reviewing material from undergraduate classes such as basic analytical techniques, physicochemical properties of chemical structures, basic applications of chemistry to the drug design process, and biotransformation processes.

In conclusion, the main advantages of using HyLighter as an instructional tool in distance education courses at the University of Florida are:

- HyLighter is an effective tool for students and instructors to transfer knowledge acquired in written modules into practical examples and authentic assessments.
- Students are able to evaluate and compare their own answers with those of their peers which provide additional discussion incentives and increased engagement.
- HyLighter enables collaborative student approaches for alternative assessment activities that complement quizzes and individual assignments.

Improving Learning Outcomes for Working Professionals

The challenges of a distance education setting have been outlined above. Although colleges and universities are increasingly investing in distance and online educational efforts, it remains challenging to evaluate the learning outcomes of such efforts in comparison to the traditional classroom setting. This is mainly based on the diverse nature of online instruction and variable integration of instructional technology.

Application of the National Science Education Standards to UF programs

The National Science Education Standards emphasize the improvement of learning outcomes through exemplary science teaching (NRC 1996). These standards embrace independent thought and development of critical-thinking skills that are often required beyond the actual knowledge acquisition process to provide students with practical skills applicable to their later professional career.

The emphasis of the UF programs in forensic sciences, pharmaceutical chemistry, and clinical toxicology are focused around fostering and encouraging students to develop and apply independent thought and critical thinking throughout the courses. This is accomplished by using a variety of tools to increase both the understanding and comprehension of students as well as to enhance learning outcomes while providing increased flexibility as compared to the traditional classroom setting.

Table 16.1 contrasts the traditional learning approach with the application and integration of the different technologies introduced in this chapter within the UF distance education programs. Furthermore it indicates how the use of these technologies improves learning outcomes.

Table 16.1. Teaching Standards Comparison Between Traditional Approaches Menioned in the National Science Education Standards and Innovations and Improvements Realized by the UF Distance Education Programs.

Traditional Approach	UF Distance Education Program	Improvement in Learning Outcomes
Treating all students alike and responding to the group as a whole	Interacting both with the whole class through discussion forums as well as providing individual support to students via e-mail and chat room meetings.	Increase student knowledge, comprehension, and evaluation through interaction and active inquiry.
Rigidly following curriculum	Provide additional resources and introduce up-to-date information relevant to the current course material, encourage students to contribute to ongoing discussions.	Application of knowledge to new scenarios, analysis and synthesis of learned content.
Focusing on student acquisition of information	Encourage independent research through individual essay assignments, use of HyLighter and CrimeSeen 360 to engage students in interaction and inquiry as a group.	Allow for application, analysis, synthesis, and evaluation of acquired knowledge through transfer to new situations and critical peer evaluation.
Asking for recitation of acquired knowledge	Engage and integrate students in the learning process—provide room for a community of learners that learn from each other, collaborative work in HyLighter, interaction through discussions.	Prepare students to become lifelong learners, increase analysis and synthesis skills essential for professional career setting.
Testing students for factual information at the end of the unit or chapter	Aside from timed individual quizzes, students have to work together as a team and with the instructor through discussions, chats, and using HyLighter to analyze and evaluate a given problem.	Engage students in peer review, critical evaluation of work, and working as a team to evaluate and solve a problem.

The current approach of the UF programs is tailored to provide students with the necessary skills to further their understanding in the respective topic while also integrating approaches that enable future scientists to apply critical-thinking skills in their workplace. The combination of individual essay assignments that evaluate each student and the peer review and interaction on discussion boards, chat sessions, and advanced tools such as HyLighter and CrimeSeen 360 provide a unique and versatile approach to advancing the knowledge, comprehension, application, analysis, synthesis, and evaluation skills of working professionals.

The significance of these improvements in learning outcomes is reflected by course evaluations in 2011 for the course PHA5433 Fundamentals of Medicinal Chemistry that are taught both online as part of the UF distance programs and in the traditional classroom setting on the UF campus (spring, summer, and fall for the online setting and fall for the on-campus setting). The online course setting also integrated HyLighter as part of the student interaction whereas the on-campus setting did not use additional technology improvements.

Another important component in an educational setting is assessment techniques that contribute to learning outcomes. Table 16.2 contrasts the traditional assessment approaches as listed in the NSES with the implemented changes for the UF distance education programs that contribute to success and improved learning outcomes of students. Assessment of progress and learning objectives is a central part at all levels of educational instruction and especially important in a distance education setting where the direct interaction between students and instructor is limited to chat sessions. Therefore, integration of various innovative assessment tools is important to provide students with a variety of pathways to applying their knowledge. Quizzes are a standard evaluation tool that can be easily delivered in a distance course. The design of the quizzes, however, can vary significantly—either based very closely on assessing knowledge directly contained in the module notes or requiring students to transfer acquired knowledge to a new setting that indicates a deeper understanding of the topic discussed. A central tool used in the UF distance education courses is essay assignments that require students to apply the knowledge learned in lessons to a question related to the topic but with the requirements of both independent research and synthesis of an opinion and scientifically justified opinion. This often requires extensive research in the scientific literature (e.g., Pubmed, Google Scholar, and so on) and extraction of essential and relevant information to answer the essay questions.

Use of the HyLighter tool further provides for assessment variation in that students provide peer review for each other and actively discuss their various outcomes in conjunction with the instructor. Self-reflection, self-assessment, and evaluation of peers is an important learning outcome and applicable for later career development. The HyLighter and CrimeSeen 360 tools together with discussion boards, chats, and e-mail communications provide for a wide range of individual and group assessment while allowing for personalized feedback that a standardized test may not provide.

Table 16.2. Assessment Standards Comparison Between Traditional Approaches Menioned in the National Science Education Standards and Innovations and Improvements Realized by the UF Distance Education Programs.

Traditional Approach	UF Distance Education Program	Improvement in Learning Outcomes
Assessing what is easily measured	Quizzes are designed to transfer knowledge learned in the course.	Applying knowledge learned in course to a new setting, apply critical thinking and expand knowledge base.
Assessing discrete knowledge	Essay assignments often require students to conduct independent research beyond the lesson notes that requires critical-thinking skills and transfer of knowledge.	Independent thinking and development of expanded view on a science topic beyond the limitations of the lecture notes.
Assessing scientific knowledge	Essay and HyLighter assignments require students to provide background information and knowledge about a topic.	Increased awareness of connectivity among science subjects and interdisciplinary approaches to acquisition and study of scientific knowledge.
Assessing to learn what students do not know	Encouraging feedback and comments from students throughout the semester, discussion board provides for an open forum in which students can interact and exchange knowledge.	Knowledge exchange and peer review as well as explanations and discussions among students provide the instructor with a better understanding of the important and challenging topics and students with a platform to increase and apply their knowledge base.
Assessing only achievement	Additional course elements (e.g., "drug of the term" for extra credit) are focusing on knowledge exchange among students while increasing the knowledge base of the class.	Stimulation of learning through inquiry and peer review that encourages class discussions and improves the individual knowledge base.
End-of-term assessments by teachers	Essay assignments, quizzes, and HyLighter assignments are consistently evaluated during the semester without a comprehensive final.	Continuous learning and inquiry while building a knowledge base through critical thinking and independent research.
Development of external assessments by measurement experts alone	Essay assignments and quizzes are developed by instructors with input from other faculty and frequent review.	Establishment of a peer review process for assessment and learning outcomes.

The course evaluations were administered online for both the distance education and the on-campus setting and included the following questions to be evaluated on a Likert scale (1: Poor; 2: Below Average: 3: Average; 4: Above Average; 5: Excellent):

1. Description of course objectives and assignments
2. Communication of ideas and information
3. Expression of expectations for performance in class
4. Availability to assist students in or out of class
5. Respect and concern for students
6. Stimulation of interest in course
7. Facilitation of learning
8. Overall assessment of instructor

Overall, 32 students in the online setting and 268 students on campus completed the questionnaire. After calculating descriptive statistics and plotting the averages and standard deviations, each question was compared between the online and on campus setting using a 2-sided independent Student t-test (Figure 16.6). Both courses received very good ratings for each question with the lowest average of 4.4 and the highest average being 4.9.

Figure 16.6. Mean and Standard Deviation for a Course Evaluation Questionnaire

Students enrolled in the online distance setting (n = 32, dark bars) are compared to students in the on-campus setting (n = 268, light bars). Statistical analysis was performed using a two-sided student t-test.

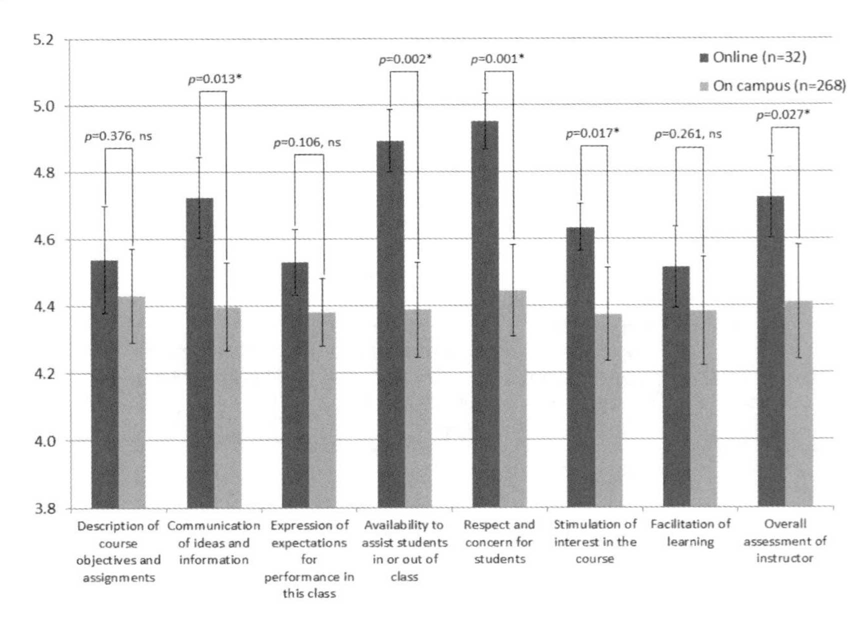

However, students in the online courses responded with higher scores to several questions. Students in the distance setting gave an average score of 4.7 for "Communication of ideas and information" compared to a score of 4.4 for on-campus students ($p=0.013$). Similarly, both "availability to assist students in or out of class" ($p=0.02$) and "respect and concern for students" ($p=0.001$) were rated significantly higher for online vs. on-campus students. Interestingly, the question "stimulation of interest in the course" was also rated significantly higher by students enrolled in the online setting vs. the traditional classroom setting ($p=0.017$). The overall rating for the instructor was also significantly higher for the online vs. on-campus setting ($p=0.027$).

These differences should be interpreted with caution, as it only reflects one year of continuous teaching for the online setting compared to one course in the regular on-campus setting. The course content is basically the same but the online environment appears to provide students with more access to interaction and communication with peers and the instructor as well as in general a higher degree of stimulation and interest in the course. Furthermore, the facilitation of learning as well as the description of course objectives and assignments was at least as effectively communicated and expressed in the online setting as in the on-campus course. These results indicate that the overall learning outcomes were as well achieved and in some regards resulted in even better outcomes in the online learning environment by implementing multiple communication and interaction tools for students to use.

Conclusions

Communication is a key component in daily interactions and especially in a learning environment. While a classroom presents opportunities to facilitate communication between the instructor and students, the learning environment in a distance education setting can at times challenge both communication and direct feedback. Over the past years, many tools have been developed that are intended to facilitate communication and interaction in a distance education setting. The integration of social networking such as Facebook, LinkedIn, or Google+ enables integration of the virtual classroom into everyday communications.

The University of Florida distance education programs in forensic sciences, pharmaceutical chemistry, and clinical toxicology use various pathways to engage students and provide instructors with tools to evaluate student knowledge and retention. To further facilitate communication, interaction, and collaboration between students and with the instructor, the social annotation platform HyLighter allows students transfer of knowledge learned during the class, application of learned material to a variety of different novel settings, and collaborative work on a project that allows both reflection and increasing abstract and critical-thinking skills.

Such interactions are crucial both for collaboration that is expected in the workplace as well as knowledge transfer beyond the theory. Students regarded the use of HyLighter as useful and applicable to the course. Though the specific example given applies to a graduate distance-education chemistry class, HyLighter can serve as a valuable tool to supplement classroom discussions in a variety of academic fields. Although several other academic institutions and instructors are using HyLighter as an additional tool for asynchronous discussion outside the classroom, its unique capabilities allow for use in both the distance education as well as traditional classroom setting.

Compared to the on-campus setting, the use of various instructional technologies such as discussion boards, chat sessions, HyLighter, quizzes, and essay assignments have indicated that the same course content can be delivered with the same or even better learning outcomes in a distance education setting. Although further research is needed, instructional technology as well as well-designed course content can improve learning outcomes and knowledge transfer in various educational settings. This is a significant advancement compared to the traditional delivery of content via lectures.

The use of HyLighter solves the key issue of evaluating learning outcomes and student processing of the encountered educational material in distance education asynchronous settings. While the individual student can first provide analysis of an abstract problem, the peer review process (after all comments are accessible to students) allows for further reflection. It further gives instructors a tool to track how students process and apply new information encountered in the class. In many disciplines, students are not able to reflect on the information but rather choose to memorize the material (Trigwell and Prosser 1991). This has often been related to time constraints and the learning environment itself. If the learning environment does not allow for development of critical-thinking skills or reflection on the learned material, students are less likely to integrate and transfer knowledge to retain it.

The integration of new technologies to facilitate communication and improve learning outcomes brings the instructor the responsibility to integrate such tools into the learning environment in a way that actually provides students with a better way to study and learn. This is not always given when it comes to social networking tools that are easily accessible yet may hinder progress because of their general nature and potential distractions. The learning environment should be separate from the casual and personal interactions but not isolated. The tools introduced in this chapter allow for integration into both synchronous and asynchronous learning environments while also providing a social component that allows students to interact.

While there have been questions about the learning environment in distance education settings, a number of evidence-based research publications that have been summarized by the U.S. Department of Education (U.S. Department of Education 2009) have shown that various forms of content delivery via distance and web-based education can provide the same learning outcomes and are equal to the outcomes of traditional classroom practices. Even with differing delivery methods (narrated PowerPoint slides, one-way or two-way video, written materials, flash animations, and so on), the various studies and the meta-analysis did not show significant differences compared to the traditional classroom setting. Any technique that can further enhance asynchronous teaching approaches and improve learning outcomes provides both students and instructors with additional resources to engage and make the distance learning environment more accommodating to institutions of education at all levels of instruction.

References

Alsharif, N. Z., and K. A. Galt. 2008. Evaluation of an instructional model to teach clinically relevant medicinal chemistry in a campus and a distance pathway. *American Journal of Pharmacy Education* 72 (2): 31.

Ericsson, K. A., R. T. Krampe, and C. Tesch-Römer. 1993. The role of deliberate practice in the acquisition of expert performance. *Psychological Review* 100 (3): 363–406.

Lave, J., and E. Wenger. 1991. *Situated learning: Legitimate peripheral participation*. Cambridge, MA: Cambridge University Press.

National Research Council (NRC). 1996. *National science education standards.* Washington, DC: National Academies Press.

Parsad, B., and L. Lewis. 2008. Distance education at degree-granting postsecondary institutions: 2006–07. In: *National center for education statistics*. Washington, DC: U.S. Department of Education.

Roper, A. R. 2007. How students develop online learning skills. *Educause Quarterly* 30 (1). *www.educause.edu/eq*

Scardamalia, M., C. Bereiter, R. S. McLean, J. Wallow, and E. Woodruff. 1989. Computer-supported intentional learning environments. *Journal of Educational Computing Research* 5 (1): 51–68.

Trigwell, K., and M. Prosser. 1991. Improving the quality of student learning: The influence of learning context and student approaches to learning on learning outcomes. *Higher Education* 22: 251–66.

U.S. Department of Education, Office of Planning, Evaluation, and Policy Development. 2009. *Evaluation of evidence-based practices in online learning: A meta-analysis and review of online learning studies.* Washington, DC: U.S. Department of Education.

Endword

Exemplary College Science Teaching: Helping or Hindering STEM Reforms?

Robert E. Yager
University of Iowa

Wow! What an exciting 16 chapters! Who would have guessed the extent of the diverse authors and the nature of their efforts to improve college science teaching? For too long there have been few who have thought beyond lectures and two-hour labs as the sole organization for college science—and usually within the boundaries of the major "disciplines." These include biology (only the recently combined botany and zoology departments), chemistry, physics, and geology. The chapters focus on nonlectures, use of local environments, case studies, study groups, jigsawing, and continuing to analyze and use technology.

This NSTA Exemplary Science Program (ESP) monograph is sponsored by the Society for College Science Teachers (SCST). We hope every member will want copies of the whole monograph series as new ideas are raised and considered. Our readers are invited to review again the words of President Brian R. Shmaefsky (in the preface). It has been typical for college science programs to merely add new science and technology information to the typical offerings. These must meet the features of major reform efforts while also attracting more to consider science careers, often including medicine, technology, and engineering. Such courses were not "typical" ones for undergraduate offerings—but few have actually discussed or planned for any changes or specific teaching alternatives to enhance student learning and more positive attitudes. It is our hope that this monograph and the new future editions planned will result in making college science teaching more attractive for all instructors *and* their students.

Jim Gee (2012) has recently called for improving college teaching in all departments and colleges. He has outlined needed changes and reported on attempts to improve instruction and provide evidence of the results in terms of their effects on student learning while also developing more positive attitudes toward the science classes. Gee has rejected the notion that the way college teaching is typically organized cannot be improved. He reports on the power and success with more student-centered courses. He has called for reforms that include specific changes in curriculum areas while arguing that different students learn differently. Successes in student groups frequently vary in specific ways. Students differ; there are students who obtain both high-end and middle-level thinking skills. On his website (*www.jamespaulgee.com*) Gee writes specifically about these differences:

> If only a few are producers and most are consumers, then a group is divided into a small number of "priests" (insiders with "special" knowledge and skills) and the "laity" (followers who use language, knowledge, and tools they do not understand deeply and cannot transform for specific contexts of use).

These differences in learners are expected and used for interpreting successes with student learning. Gee recommends "post-mindless progressivism" and has offered ways that both students and teachers can be described in situations that are designed to provide more effective learning environments. He has proposed 17 features of environments needed for achieving real learning and the attainment of a truly student-centered curriculum. The authors of our 16 exemplars have approached many of these features, but are only beginning to alter total science curricula in their universities. Gee's suggestions include actions to attain the following:

1. Multiple routes to full and central participation for all members of a group, a group organized around an interest and a passion to which the interest might lead.
2. Multiple routes to everyone learning to produce the knowledge, dispositions, skills, and tools necessary to sustain, extend, and transform the interest and the passion.
3. Interest kindles motivation and the desire to explore. The interest must then be channeled into a passion so that learners persist towards mastery via a great many hours of practice. Otherwise learners need to find another interest that will lead to a passion.
4. Learning is well designed so that learners are immersed in well-structured, well-designed, well-mentored, and well-ordered problem solving inside experiences where goals are clear and action of some sort must be taken.
5. Feedback is copious. Lots of data on multiple variables across time is collected and used to resource learners; assess their growth and development over time; and assess, compare, and contrast (for both learners and stakeholders) different possible trajectories to master, including ones that lead to innovation and creativity.
6. Learning and assessment are so tightly integrated that finishing a level of learning is a guarantee of mastery at the level, a guarantee that learners can solve problems and not just retain facts (but use facts as tools for problem solving), and a guarantee that learners are well prepared for future learning.
7. All learners must master one or more specialties at a deep level, be able to teach that specialty to others, and be able to learn new things when needed from others.
8. All learners must be able to pool their specialty with other people's different specialties and integrate their specialty with other people's specialties by seeing the "big picture" to be able to solve problems that no one specialty can solve.
9. All learners are well mentored by "teachers" and peers at various levels, as well as by the presence of smart tools and well-designed problem solving environments (both real and virtual). All learners must learn to mentor.

10. "Teachers" are designers of learning environments that meet all the above conditions and they resource people's learning in an adaptive and contextually responsive way.
11. Direct instruction and texts are offered "just in time" (when learners can put them to use and see what they really mean) or "on demand" (when learners feel a need for large amounts of instruction or text in their trajectory of problem solving).
12. Failure is used as a learning device, so the price of failure is, at least initially, kept low so all learners are encouraged to explore, take risks, and try different learning styles.
13. Learners are shown through modeling and made well aware of adult or professional norms for the skills and dispositions they are developing and held to high standards based on these norms in ways that make clear every learner can reach those norms should they choose to put in the time and effort.
14. Learners come to see and be able to use the relationships and connections among different types of skills and knowledge, often "stored" in different people, as well as to understand the larger social, environmental, and cultural implications of any proposed solution to a problem.
15. Learners can integrate and see the connections among science, mathematics, social science, the humanities, ethics, and civic participation. In today's world this often means seeing how the same social and digital tools can be used for different types of discovery and interventions in the world across the arts, sciences, and humanities.
16. Learners are well prepared to learn new things, make good choices, and be able to create good learning environments for themselves and others across a lifetime of learning.
17. All learners are well prepared to be active, thoughtful, engaged members of the public sphere (this is the ultimate purpose of "public" education), which means an allegiance to argument and evidence over ideology and force and the ability to take and engage with multiple perspectives based on people's diverse life experiences defined not just in terms of race, class, and gender, but also in terms of the myriad of differences that constitute the uniqueness of each person and the multitude of different social and cultural allegiances all of us have. (Gee 2012)

Another reference is suggested for readers and authors to assist them with improving college teaching. It is entitled *The Learner-Centered Curriculum: Design and Implementation authored* by Cullen, Harris, and Hill. It includes 200 important and useful references, mostly of research and study groups that define teaching as more student-centered. The authors indicate both logic and know-how for many that experience learner-centered curricula. These exemplify the push for change while also supporting new ideas to try. This new book explores ways for improving student learning in all college science areas. It suggests new ways that teachers need to explain the importance of learning if real progress with student learning is to be achieved.

Although learner-centered practices are being employed with great success by many teachers in individual classrooms, not much attention has been paid to how those experiences could or should be linked together. These efforts need to be supported through curriculum design. Curriculum is the heart of what we do. It needs to be examined and be aligned with learner-centered practices if our institutions (especially colleges) are going to become truly learner-centered. We need to examine curriculum design from the vantage point of postmodern and learner-centered perspectives.

We need to begin by posing the question, *Why do we need to redesign the curriculum?* The answer that was offered was that learner-centered curricula should provide opportunities for our graduates to become creative, autonomous learners, and the people most needed for the 21st-century work force. We need to examine curriculum as it manifests itself in the modern era. This means changing the instructional paradigm in order to identify assumptions about curriculum that have been accepted as facts, or unexamined realities of the paradigm (the one traditionally accepted and practiced). New ways of looking at curriculum are needed, and in a postmodern way that challenges assumptions predicated on a modern view of learning and education. Recognizing that implementing the kinds of changes will be challenging, strategies were offered by Cullen and colleagues for implementation. They suggest rhetorical strategies for leaders to employ that itself are postmodern by design. To illustrate that these postmodern, learner-centered design principles are not just theoretical constructs but very real options, examples were offered to provide specific curricula that vary in degree of learner-centeredness. A rubric was offered for use in assessing the degree of learner-centeredness characterizing a curriculum. The research offered analyzes practical ways of supporting learner-centered curricula, through assessment practices, use of technology, and consideration of physical space.

It is argued that the curriculum, whether instructional or programmatic, needs to be redesigned. Individual instructors employing learner-centered practices in their individual classrooms have greatly increased understanding of what we mean by "learner-centered." But for the paradigmatic shift to be fully realized, all need to look beyond what takes place in individual classrooms and consider how those experiences link to and inform one another.

The 21st century is dependent on autonomous learners who can be creative and innovative. Students' minds need to be cultivated to meet the future needs. Students must be trained in *multiple* disciplines. They must be creative, for individuals without creative capacities will be replaced by computers and will drive away those who do have the creative spark (Gardner 2008). It is believed that the research that informs the learner-centered agenda offers the opportunity to create learning environments to foster these qualities in learners—all by placing learning and the learner at the center of focus. To quote Gardner: "The survival and thriving of our species will depend on our nurturing of potentials that are distinctly human. It is the focus on learning as a distinctly human activity that characterizes learner-centeredness."

It is important to envision catalysts for thinking about curriculum design for all K–16 science offerings. Real learning is a result of continuous discussion and reflection. This is how the process of curriculum redesign should also begin. It surely is what has been offered here in our monograph. Some of the approaches by the authors provide needed conversations to stimulate the process further. We expect this monograph to encourage STEM reforms while also stimulating more to try!

References

Cullen, R., M. Harris, and R. R. Hill. 2012. *The learner-centered curriculum: Design and implementation.* San Francisco: Jossey-Bass.

Gardner, H. 2008. *Five minds for the future.* Boston: Harvard Business Press.

Gee, J. 2001. Progressivism, critique, and socially situated minds. In *The fate of progressive language policies and practices,* ed. C. Dudley-Marling and C. Edelsky, 31–58. Urbana, IL: NCTE.

Gee, J. 2012. Beyond mindless progressivism. *www.jamespaulgee.com/node/51.*

Contributors

Elizabeth Allan, author of "Revising Majors Biology: A Departmental Journey," is an associate professor of biology at the University of Central Oklahoma, Edmond, Oklahoma.

Sandhya N. Baviskar, author of "Implementing Jigsaw Technique to Enhance Learning in an Upper Level Cell and Molecular Biology Course," is an assistant professor of cell biology and science education at the University of Arkansas-Fort Smith, Fort Smith, Arkansas.

Grace Eason, author of "Service Learning in an Undergraduate Introductory Environmental Science Course: Getting Students Involved With the Campus Community," is an associate professor of science and science education at The University of Maine Farmington, Farmington, Maine.

Katherine B. Follette, author of "The Road to Becoming an Exemplary College Science Teacher," is a graduate student in astronomy at the University of Arizona and an adjunct instructor of astronomy at Pima Community College in Tucson, Arizona.

Oliver Grundmann, coauthor of "Graduate Distance Education Programs in Forensic Sciences, Pharmaceutical Chemistry, and Clinical Toxicology for Working Professionals: An Evolving Concept With Practical Applications," is a clinical assistant professor at the University of Florida, College of Pharmacy, in Gainesville, Florida.

Beth Ann Krueger, author of "Take Your Students Outside: Success with Science Outdoors," is a professor of science at the Central Arizona College–Aravaipa Campus, Winkelman, Arizona.

David G. Lebow, coauthor of "Gradate Distance Education Programs in Forensic Sciences, Pharmaceutical Chemistry, and Clinical Toxicology for Working Professionals: An Evolving Concept with Practical Applications," is the CEO of HyLighter LLC in Tallahassee, Florida.

Thomas R. Lord, author of "Requiring College Students in a Plant Science Course to Take Control of Their Learning and Students Teaching Students: Jigsawing Through an Environmental Biology Course," is a professor of biology at the Indiana University of Pennsylvania, Indiana, Pennsylvania.

Teddie Phillipson-Mower, author of "The Great Debate: Revising an Old Strategy With New Frameworks," is the director of the center for environmental education at the University of Louisville, Louisville, Kentucky.

Joseph Salvatore, coauthor of "Peer-Led Study Groups as Learning Communities in the Natural Sciences," is an assistant director of the Science Learning Center at the University of Michigan, Ann Arbor, Michigan.

Claire Sandler, coauthor of "Peer-Led Study Groups as Learning Communities in the Natural Sciences," is a director at the science Learning Center at the University of Michigan, Ann Arbor, Michigan.

David N. Steer, author of "Clickers in the Geoscience Classrooms: Pedagogical and Practical Considerations," is a professor in the department of geology and environmental science at the University of Akron, Akron, Ohio.

Thayne L. Sweeten, author of "Assessing Student Attitudes and Behaviors in Interactive Video Conference Versus Face-to Face Classes: An Instructor's Inquiry," is a lecturer in the department of biology at Utah State University, Brigham City, Utah.

Linda L. Tichenor, author of "Assessing Learning Outcomes of the Case StudyTeaching Method," is an associate professor at the University of Arkansas –Fort Smith, Fort Smith, Arkansas.

Holly J. Travis, author of "The Student-Centered Lecture: Incorporating Inquiry in Large Group Settings," is an assistant professor of biology at the Indiana University of Pennsylvania, Indiana, Pennsylvania.

Bonnie S. Wood, author of "Lecture-Free College Science Teaching: A Learning Partnership," is a professor of biology at the University of Maine at Presque Isle, Presque Isle, Maine.

Ellen H. Yerger, author of "Student Perspectives on Introductory Biology Labs Designed to Develop Relevant Skills and Core Competencies," is an assistant professor of biology at the Indiana University of Pennsylvania, Indiana Pennsylvania.

Index

Page numbers printed in **boldface** *type refer to figures or tables.*

M

N

O

P